Vinyl Records

Valuable Information for Novice Record Collectors to Buy

(The Ultimate Beginner's Guide on How to Get Started With Vinyl Records)

Boyd Dodson

Published By **Oliver Leish**

Boyd Dodson

Vinyl Records: Valuable Information for Novice Record Collectors to Buy (The Ultimate Beginner's Guide on How to Get Started With Vinyl Records)

ISBN 978-1-7779883-4-0

No part of this guidebook shall be reproduced in any form without permission in writing from the publisher except in the case of brief quotations embodied in critical articles or reviews.

Legal & Disclaimer

The information contained in this book is not designed to replace or take the place of any form of medicine or professional medical advice. The information in this book has been provided for educational & entertainment purposes only.

The information contained in this book has been compiled from sources deemed reliable, and it is accurate to the best of the Author's knowledge; however, the Author cannot guarantee its accuracy and validity and cannot be held liable for any errors or omissions. Changes are periodically made to this book. You must consult your doctor or get professional medical advice before using any of the suggested remedies, techniques, or information in this book.

Table Of Contents

Chapter 1: For The Record 1

Chapter 2: Huge Market For Records 12

Chapter 3: For The Record 24

Chapter 4: How To Sell Your Records In
Classified Ads 38

Chapter 5: Take A Listen 61

Chapter 6: The More Stores The Better . 72

Chapter 7: Shipping Time If By Means Of
Using Mail 85

Chapter 8: For The Record 95

Chapter 9: The Vinyl Revival................. 114

Chapter 10: Structure Of A Vinyl File.... 126

Chapter 11: The Process Of Vinyl Record
Production 135

Chapter 12: Groove Geometry............. 147

Chapter 13: Cartridge Styles And Types 161

Chapter 14: Different Sorts Of Cartridges
.... 173

Chapter 1: For The Record

Vinyl facts sell for a blended motive. In order of importance:

1)   Condition

Buyers want statistics in super condition. I don't want to look like a damaged document myself however scratches on vinyl decreases the price of a data substantially. For example an real reproduction of Pink Floyd Dark Side of the Moon sells for $30 US minimum. I'm talking approximately a NEAR MINT grade reproduction. If your reproduction has scratches and the grade is most effective GOOD, your vinyl would best promote for $15 maximum. Do you word what I'm talking about? So, when you have some particular NEAR MINT situation copies of information, you're wearing a few luxurious stock.

HOW DO YOU BECOME A EXPERT ON GRADING THE CONDITION OF A RECORD- At first as soon as I commenced out I requested a my buddy at ther used record to appearance over my information and provide mea grade for each one. After some reviews I have become an professional at grading my very own records.

2) Artist

There are positive well-known artists that hold to promote thoroughly and continuously will. The Beatles as an instance are one of the top artists that preserve to sell properly in used records stores. Everytime I visit used record stores it looks like there are continually small portions of those artists available. Bands from the 60's and 70's are pinnacle promoting artists as nicely and when you have Vinyl from these some years you could make some specific money. When you start selling take a look at your promoting with well-known artists and be aware which sells the extraordinary.

HOW DO RECOGNIZE RECORDS AS AN ART PIECE. – Generally the most colorful Vinyl covers from the 60's and 70's promote thoroughly for me. Many of my clients have requested me for Pink Floyd record covers as they've alot of revolutionary fee. TAKE OUT OR MOVE

three)   Rareness

Record clients love uncommon data. Whether you're promoting online, at Record shops or Flea markets, make sure you be aware to your advertisements that the file is RARE. How will you're privy to it's uncommon? Google it or appearance it up on Record Collectors mag. Try list it on Ebay and spot how many watchers you get. Make certain it's the primary data you point out at the same time as you sell your report! Rare facts sell thoroughly!

HOW DO YOU KNOW THAT A RECORD IS RARE – Check on line and Google the call of the document. If you don't select up sufficient information try and appearance on Ebay and appearance what the report is selling for. Make certain which you healthful the code on the duvet of the report or inner of the report label. Don't anticipate that clearly because the quilt picture suits that the report listed that the record is unusual is unusual. It may be a more modern release therefore the rate is probably plenty a good deal much less.

four)   Interest or Music flavor

There are lenders sincerely interested by exceptional varieties of track. They normally purchase that fashion of tune. They may be severe Rock or Jazz creditors. Try to 0 in in this marketplace whilst you start to promote facts. See what sells in particular sorts of track and purchase and promote your inventory based totally definitely on this.

HOW DO YOU TAP INTO INTEREST OR MUSIC TASTE WHEN SELLING RECORDS-

Keep an splendid sort of one in each of a kind styles of music reachable. Once you srart selling hold tabs on what style sell the excellent. Then begin to adjust your inventory to the tastes of your clients.

SIDE ONE Song three

My History of Success

I've been selling vinyl records for pretty a few years now. I commenced decrease decrease returned in early 2002. My inventory of vinyl statistics got here from from purchasing for records each week once I became more youthful and locating vintage facts left from my brothers. I furthermore traded information with friends and companion and children and constructed up pretty a hard and fast.

The great type of vinyl facts I had in my collection come to be approximately 500. One day I got here at some point of an advert inside the newspaper which stated "Used Vinyl Records Wanted! I'll Pay Top

Cash." Well I commonly want to cash in on a superb opportunity so I determined to promote off a group of my vinyl to this used document company that changed into based in my region. Beatles, Led Zeppelin, Queen , Kiss information all were supplied to this used file dealer. I bear in thoughts he gave me $two hundred. "Wow", I stated that is remarkable, what an smooth way to make coins. I didn't understand at that issue that a prime a part of my series disappeared for in reality a small part of what I might also need to've were given. Shame on me!

I located the tough manner and that during case you don't do your homework you can leave out out on a bigger pay day. This is proper for everything in existence. I've made many mistakes collectively with selling statistics way below their suitable cost. I took a few years to build up my stock over again. I turned into purchasing for at report suggests, used record shops and started to buy online. I positioned that I may additionally want to shop for any records from any a part of the arena if I preferred by using using the usage of the internet.

If there may be one scratch or scuff at the vinyl or cowl your record may additionally want to lose 50 to 75% of its fee!

In 2004 I commenced promoting my information once more to the used document stores once more but this time I started out out making better cash. I located report magazines which consist of Record Collector   and locating the fee of my stock. I became surprised to appearance how a good deal my information were absolutely nicely well worth. I saved searching for an increasing number of records and it wasn't till early 2008 that I commenced out selling on line. This became as soon as I began out make a intense income. I changed into selling then looking for, and generating some massive earnings. I used internet web sites which includes Ebay, Ebid, and Amazon. I didn't open my private website online as I'm on top of things with making my personal internet net internet page. I then commenced looking for unusual information, shopping for and flipping them over for a profits. I like doing this as it doesn't absorb an excessive amount of of my time up and it's clearly easy to do.

I list lots of information consistent with yr on-line and make proper passive income doing this. But that's now not all. I regardless of the truth that use many other venues to promote my data.

You can promote at:

- FLEA MARKETS

- ON LINE – MULTIPLE SITES

- RECORD MAGAZINES

- CLASSIFIED ADS

- USED RECORD SHOPS

What about Vinyl file shopping for and promoting. Well it maintains me busy and I love doing it. I choice you could have a look at from my reviews and take the identical steps I sincerely have found. It works for me and I'm tremendous it's going to offer you the consequences you need.

If there may be one most important detail that I have observed out from promoting Vinyl facts is that a document wishes to be outstanding to keep to its rate. If there can be one scratch or scuff at the vinyl or cowl your report want to lose 50 to 75% of its fee!

SIDE ONE – Song three

Selling records makes experience. Why?

- There is a big market for vinyl statistics

- The hobby maintains to increase

- Its an easy product to distribute

- Your inventory expenses are exceedingly low

- You don't need quite a few region to keep data

- The assets for getting are big

- You can do it part time.

- The profit margin is high

- You can sell multiple strategies

FORMAT	MARGINS	SHIPPING	DEMAND
Vinyl Records	High	Low fee	High
CDS			
Vintage Mags			
Lower	Low price	Lower	

Comics, CDs, antique magazines have visible a decline in income in the last few years. You should buy CD for simply pennies on Ebay. CD Collectors have misplaced interest over the previous few years. Records profits but hold to increase. In truth some tune shops have reopened up vinyl earnings. The enlargement of song into the internet moved vinyl information into the antique beauty. Don't be amazed in case you begin to see new vinyl file shops pop up on your network inside the destiny.

Vinyl file albums' earnings maintain developing. Quickly shifting up in income an entire lot as 15 percentage up in 2007 and a considerably 89% greater in 2008

CD SALES HAVE DROPPED

Chapter 2: Huge Market For Records

While record sales have seen outstanding improvement in profits, CD sales preserve to drop. This is due to fact track is available online at any time. There is not any cause to acquire CD music as there isn't the charge seen in Vinyl LP or Vinyl single statistics. The nostalgic rate isn't there. LP vinyl information continue to be greater of a cult object than CDs or Digital Music.. The name for created through using the Vinyl format is particularly more appealing than the virtual CD format. You'll be aware at the same time as you appearance online, a CD promoting on Ebay is miles less expensive than what a file sells for. There is more of a music bundle than there can be with CDs. The item is greater and loved as paintings piece as nicely.

Radiohead, Beach boys and Jimi Hendrix have had re-liberating of antique facts onto new Vinyl. This craze started in 2008 and loads of various artists repeated this same concept. This makes used vinyl plenty more appealing as creditors begin to search for the authentic variations of those re-releases.

THE INTEREST CONTINUES TO GROW

Best Buy have even started out promoting vinyl information as this organisation has seen as extremely good addition to beautify sales in their shops. The plan is that Best buy will begin piloting 100 stores for Vinyl record profits in 2010 and 2011. Wallmart is also deliberating selling new vinyl facts of their stores proper away. So what does this will will let you recognize?. It's time to get into this marketplace and start your organisation selling New and Used data proper away.

AN ORIGINAL COPY FROM FROM THE SAME ARTIST MIGHT FETCH YOU DOUBLE THE PRICE OF A RE-RELEASE

WHY SELL RECORDS

I changed into very thrilled to look the amount of used file shops which have popped up in my region over the previous couple of years. A accomplice of mine who had a completely small shop in my network extended to a facility double the size due to fact his earnings and speak to for is developing. He resources to public including myself and gives his vinyl on-line as nicely. This is a key sign that more and more humans are starting to realise that that could be a first-rate business organization to be in. You need to be aware of this as properly.

As I mentioned artists are starting to launch their music in CD layout and vinyl for extras income. The idea proper right here is to create a lenders object that their lovers will buy. The price of vinyl facts can double and triple the fees of CDs. Some new vinyl information cost around $30 which makes the used vinyl a whole lot more attractive to shop for. Here's wherein you can coins in. A ultra-modern vinyl file remake no matter the fact which could promote thoroughly if you turn round and sell it weeks later. An specific replica from the same artist would possibly likely fetch you double the rate of a modern launch.

ITS AN EASY PRODUCT TO DISTRIBUTE

A used vinyl record is relatively cheap to deliver and get hold of. I purchase my information in bulk on line and store tons of cash in shipping fees. I order 5 to ten records at a time and I normally recommend that my customers do the identical to keep delivery prices. Records can deliver anywhere from 3 to eight dollars US relying on their weight thickness. 45pms singles supply for 1/2 of the fee. The rate of packaging is also very reasonably-priced. You should purchase unique shipping cartons to supply facts for a vey low charge on Ebay. I make my very very own shippers from cardboard I accumulate from neighborhood industries. Just preserve in thoughts to supply your statistics with constant packaging otherwise you'll pay attention out of your clients. I'll cover this in later chapters.

YOUR INVENTORY COSTS ARE RELATIVELY LOW

Depending on how large you want your document agency to be, your stock price have to be as an opportunity low. As I cited previously, seeking out information in bulk can be very smart to do to maintain costs. By shopping for in bulk you keep your inventory value low. Your common charge of purchasing mainstream statistics have to be approximately $3 in line with vinyl document and $1 constant with 7in. I purchase facts from Music Stack for $four for a Near Mint replica and flip round to sell it for $12. 30 % of your stock have to the better priced vinyl which you promote from $20 to $30. You additionally should purchase better priced object in bulk if you find out a accurate provider.

HOW DO YOU KEEP YOUR INVENOTRY COTS LOW – Turn over your records as regularly as possible. By promoting through distinction venues you could go along with the drift you inventory. Don't fear about losing cash on a few gives. It will all stability out right right into a profits if you rate your and buy your statistics successfully. You go along with the float your invenotery alsdo via promoting at a reasonale fee. Don't overstate your costs too excessive o you'll get stuck with an excessive amount of stock

YOU DON'T NEED A LOT OF SPACE

I ran my report commercial business company from my basement . It took up a one small nook of my place. I generally saved them in big plastic sturdy crates. I did this so I might also moreover want to pick out out up my stock at any time to show off it at Record stores or Flea Markets.

HOW DO YOU SET UP SPACE TO STORE YOUR RECORDS- Your facts want to be stored in a well lit, dry area of your home. Mant sellers save statistics in bloodless basements onlty to famous the cardboard on report covers emerge as mouldy or dusty. Keep a phase aside wherein youcan get at them resultseasily and watch how you stack the bins. Try now not to combine up your record vicinity with exclusive devices within the house and this could cause an twist of fate with an item falling onto a disc and inflicting damage to the record

Quite the numerous document shops I deal with are very small as records take in very little area and are easy to shop. Of course now the best inventory I carry is facts that I need to preserve in my prize possessions. My suppliers ship all my statistics and bundle deal to my customers for me. I'll describe the manner to do this later. Overhead prices are minimum. You want a computer of course, printer and that's it. Show percent

TRY TO STICK TO ONE SOURCE FOR BUYING RECORDS. WHEN YOU DEVELOP A STRONG BOND, THE SELLER WILL GO OUT OF THEIR WAY TO GET YOUR SHIPPED TO YOU

THE SOURCES FOR BUYING RECORDS ARE HUGE

You can buy facts from flea markets however the fable is maximum of the time they're underneath top notch and tough to resell. This isn't real. I determined terrific wonderful information at flea markets. I normally preserve on with a flea marketplace company that has masses of inventory.

I moreover purchase my statistics from Music Stack. You can discover a few element there at a low price. There about 1 million facts available to buy from Music Stack. Gem data is a few exclusive large outlet . They have a great comments tool so it's smooth with the intention to display your supplier.

Ebay sells masses of information. I normally stick with a person that operates a Ebay hold and has a massive inventory of product.

HOW DO YOU FIND A RELIABLE SOURCE FOR BUYING RECORDS – Start trying out the difference companies that I indexed above.. Order big quantities and keep tune of notable and transport time. Once you land a reliable supplier it makes your machine a lot much less hard. Eventually, once you have got were given self perception in them, they're capable of do all the artwork of packaging and delivery for you.

THERE ARE ABOUT 1 MILLION RECORDS AVAILABLE TO BUY FROM MUSIC STACK.

YOU CAN DO THIS PART TIME

The splendor approximately selling records is that you can do it factor time and while your playing your preferred music. I started this way and notwithstanding the reality that I spend I extraordinary deal of time on it, I consider it a detail time hobby because I love music and statistics. Many of the stores I address are open for simplest part of the day and stay open on weekends as most of their commercial employer comes through right now. Due to reality the internet is a number one hassle in selling records, maximum of the dealers I realize have transformed to the net and make excellent passive earnings promoting Vinyl. Start small and don't give up your fulltime manner. Spend some hours each days building your enterprise.

Here's the number one motive you have to bear in mind promoting statistics.

PROFIT My commonplace charge for a file I supplied on-line is $12 to $15 greenbacks. My fee to buy emerge as normally simplest $four dollars. That's a big earnings. I truly have dealers that I've worked with for some years and that they promote me information at a wholesale fee. Try to expand a high-quality walking dating together with your providers. They will do a little thing for you including rush shipments or drop transport. My report sellers were extraordinary and are a primary purpose for my success.

Of path there can be furthermore the opportunity to resell the outstanding unusual report at a much better charge. Remember there are enthusiasts of music with the intention to pay dearly for information. You simply need to pull them into your save. Don't get discouraged in case you're making enough earnings. Once you get going the profits will begin to roll in. You need to preserve at it. Continue to promote in as many one-of-a-type venues and net websites as possible. Some dealers I realize only hhave a earnings margin of 50 cents to as a minimum one greenback but there amount of income are fairly immoderate. They sell their record very reasonably-priced and make small income profits at the same time as promoting hundredsof facts in line with month. There are a few dealers that only recognition on promoting highly-priced rare records even as promoting at a far better earnings margin. You determine what works excellent for you.

Get Going Selling Records

Your stock is each crafted from information you've had lying spherical or they'll be devices your purchases. Hopefully you have got got an inventory of as a minimum two hundred to 500 minimum to start.

Later on this e-book I'll offer an reason for the diverse techniques you can find stock.

The first element you need to is type your records. I normally kind them alphabetically for a begin. Then I label them. I'll communicate similarly in this a chunk later I the e-book.

Chapter 3: For The Record

The huge aspect approximately promoting information is as you sell you update. What I imply thru that is even as you sell a duplicate of your document proper now update it with a few special replica of the equal urgent. Let's face it, in case you've offered it as soon as, you can sell it again. Also when you examine the information that promote well continuously supply them to your stock.

Copy what the a success dealers are doing. This gets you going very speedy. Look at their commercials, what facts they may be promoting and wherein they're promoting. If they can be a success so are you capable of.

Have one thing and one element best higher than absolutely everyone else to offer. I've determined out that you could have a one up on everyone you want via supplying some element appealing. Free transport,

10% off. Let them apprehend you could find any most any file that they want to shop for.

Also once you've supplied a record in your purchaser don't allow them to circulate. Offer them every other better priced record.

I might also moreover appear to be a damaged file... pardon the pun however I'll say it again and again. Sell in as many locations as possible. In file stores, net, community advertisements, DJ's, flea markets or maybe vintage stores. Populate your statistics into as many venues as feasible as you'll have extremely good fulfillment. I'll offer you with more examples of this in a while on this e book.

SIDE ONE – Song four EMERGING SALES TRENDS FOR VINYL RECORDS

What shape of vinyl records promote?

The interest in Vinyl facts continues to develop and there are one-of-a-type types of records that sell for better prices. Certain varieties of musical recordings supply excessive profits fees. Jazz statistics and particular Broadway solid and film soundtracks tend to provide a more lively market and further charge. Also, early rhythm and blues information and the doo wop sound are especially valued and collectible. Among classical statistics, the most precious are orchestral performances, then solo instrumental, chamber tune, concertos, solo vocal operatic arias, and subsequently, complete operas. To a few lenders, whether a record is mono or stereo affects the fee. See the Tips under for rising tendencies.

Rare – Examples are records which is probably limited urgent. There are artists that have released limited pressing in one-of-a-kind countries to test their income. Sometime bands ought to come out with limited collector pressings for his or her fanatics. If you may find out a limited urgent you could word they sell for a steep fee.

Japan troubles cross for some of coins as well. I in recent times bought a Japan Dark Side of the Moon for $100. I furnished it for $70. There always may be hobby in rare and uncommon records Beatles and Elvis. If you've got records with posters, lyric sleeves, calenders, or photographs sell these records at a far better price. Many creditors need records with added proper particular covered devices. Let me provide you with an instance of a few rare statistics which have bought for huge coins. I absolutely choice that you have are able to discover some of the ones on your collections or are capable of land a few to your searches:

• Beatles – Please Please Me- Parlophone UK Gold Black Label -1st Stereo - SOLD $five,600

• The Del Tours -45RPM – Bling Girl/ Sweet and lovely 45RPM - SOLD – $4,0.Five

- A Rare Stones Promotional LP - SOLD $nine,two hundred

- Dead Weather – Alternative Rock – SOLD for $3,two hundred

Remember some of those record listed above aren't clearly too old or in antique

Category. They are unusual and difficult to find out tune LPs or forty five's. These have been presented sooner or later of on-line auctions. Collectors preserve to pay pinnacle dollar for uncommon record reveals.

Deceased Artist or band with Dead Artists – I are aware about it sounds uncommon but the fee of information can flow up overnight if an artist passes away. Nirvanna data sell properly simplest due to deceased Kirk Cobain. Michael Jackson vinyl statistics took off after he handed manner some promoting for $60 to $eighty. In the previous couple of years after 2008, a marketplace sprung up for rock records from deceased cult music starts offevolved offevolved at the side of Janis Joplin Jimi Hendrix, Jim Morrison. Doors records have supplied extremely properly for me.

Collectors of 45s – Theres a whole lot of pastime trading and selling 45's. Early rock stars and 1950 rhythm and blues are warmness proper now. There is tremendous hobby in uncommon and uncommon facts (overseas problems, and so on) from Elvis and the Beatles. However, most of their facts have little value (because of the fact such lots of copies have been produced). In distinct terms, they have been all the identical. There is also vast hobby in facts that got here with enclosures, which consist of posters of the artists.

Popular – These maintain to promote properly. The huge bands of the 60's, 70's 80's hold doing well on the vinyl charts. One of the pleasant promoting vinyl records over the previous few years have come to be The Beatles Abbey Road. You can find out many copies of Abbey Road on Music Stack or Gemm Records.

Records from the 60s- They live in excessive name for. The covers from this era are crafted from strong thicker cardboard and have a wonderful charge. The vinyl is likewise thicker rigid and more of a collectors dream. Collectors additionally body the ones covers or vinyl.

Punk- You should purchase punk data at used vinyl shops and sell them on Ebay and make proper cash. There is immoderate call for for punk records because of truth most of them were produced in the short period of 1977 to 1980 and there's now not hundreds of them spherical as compared to mainstream track.

Jazz – This style of song is huge at the internet. Jazz facts, Broadway originals and soundtracks to movie sell very well 12 months round. I enormously sold the Soundtrack to Rocky IV for $25 to Someone in the Southern USA. Soundtrack data and Jazz due to popularity have first-rate resale rate. There are dealers that deal strictly in Jazz music. There are quite a few Jazz lovers accessible if you're taken into consideration one in all them and understand your artists you can sell this form of song on your customers.

Latin- Lovers of Latin tune mainly from USA have a massive following. I see a super form of Latin tune dealers round and there sales are remarkable

Classical information – Collectors here are searching older orchestral performances and moreover ask for solo instrumental and chamber music. I've seen huge vinyl income for vocal and operatic arias, and complete operas.

Mono or stereo- Does it affect the price? – Many of my clients ask for Mono records despite the fact that Stereo sound plenty better regardless of the fact that there are positive Mono recordings that sound tremendous as nicely. If you have a terrific deal of Mono information attempt promoting them at a better mark up.

SIDE ONE – Song five

Where to sell facts.

If your going to sell to used Vinyl shops installation a time for the proprietor to put apart at the way to look at every considered one among your facts. Try no longer to sell your whole collection proper now. Attempt to connect with more than one provider to promote and buy information from. If it's Ebay or Amazon as an example take a look at in with each this form of internet web sites so that it will start listing your statistics.

FOR THE RECORD

Look over this segment cautiously. There are many possibilities and masses of stores wherein to promote your records. If you supplied your facts in definitely every person of these regions you may make a top notch earnings doing so.

Brick and Mortar

You can open a small keep inside the city. A lot of my friends have unfolded very small stores inside the metropolis. .

I observed on weekends those small stores are full of might be customers. A few of these men have moreover stepped forward into larger shops because the selection for grows. The project right proper right here is keeping up with the condominium fees. Consider asking your close by used ebook maintain for some cheap condominium vicinity inside the lower again. Books and records pass along very well.

HOW TO OPEN A BRICK and MORTAR STORE- The quantity of available shop area has grown over the previous couple of years. A lot of corporations in my place and in some additives of america have close down. This presents a very good possibility to ask for a lower apartment price. You also can ask for a quick time period rent truly to get your commercial agency started out and then flow from there. When you use a shop you may need a solid plan to cover your expensives which incorporates a complete advertising plan to sell your keep. The advantage of a brick and mortar maintain is that during case you put it up inside the right area, it can be a goldmine.

On the WEB

There is a big possibility to promote statistics at the Web. Your facts are open on the market to international. I might probable advocate whilst you start promoting you stock promote on Ebay first. Then begin selling on them Amazon. Each of these net sites have plenty and hundreds of capability clients and if you set up your sale advertisements efficaciously you can make plenty of cash. If you've got a as a substitute massive inventory of vinyl data to begin attempt installing some element on Music Stack or Gemm information. You can also begin your non-public internet site online and promote your statistics right here. You need to recognize what you're doing right here and recognise how to sell your net web page otherwise maintain on with Ebay or Amazon. If you're confident approximately putting in your record store via a Web keep start with Go Daddy. They have low rate websites to start with. Vinyl dealers every so often have 10 to fifteen websites going at one time and rake in first rate earnings. If you're assured you've got sufficient inventory or a manner of obtaining loads of inventory that is a positive manner of being a a success file dealer.

FOR THE RECORD

I WOULD SUGGEST YOU START SELLING YOUR RECORDS ON EBAY AND AMAZON. EACH OF THESE SITES HAVE MILLIONS OF POTENTIAL CUSTOMERS

Flea Markets

There is a flea marketplace near my home that sells records and CDs this is open most effective on weekends. The location is complete of people that gobble up statistics hourly. You can set up a income area at a very low charge. The best small downside is your margins may be honestly decrease because of truth humans a flea market expect a low charge object. But don't let this prevent you from selling at Flea Markets. The net website traffic at Flea Markets is big on holidays and weekends and also you'll discover that it's a notable way to promote off your inventory. Also be conscious that clients at flea Markets will need to bargain lots so be prepared. I determined that information that I couldn't promote on the Web or in a shop presented higher at flea markets.

HOW TO BE SUCCESFUL SELLING RECORDS AT FLEA MARKETS:

1-Keep turntables at your cubicles to your clients to try the statistics. This offers them self assurance that they may be purchasing for a outstanding terrific file with the aid of trying out the disc first.

2- Keep your Showcase facts excessive at the wall display clearly all of us on foot by means of using using can see them

three-Play music load in case you're allowed to do that. Give your Flea Market preserve a party environment.

4- Keep it prepared and easy

five –Enroll the purchaser right right into a month-to-month on line month-to-month guide to preserve them butying.

Chapter 4: How To Sell Your Records In Classified Ads

1 – Place an advert in close by on line web sites or in neighborhood papers. Don't pay for those advertisements. Use the loose commercials as there are penty of them.

2 – Offer to sell your statistics in bunches of 10 to twenty if viable. Make fantastic there may be little room to barter fee.

three –Mention inside the advert it's a short time provide of 3 days. This receives the purchaser to behave speedy.

four- Try to sell a good deal less costly lower grade records that you may have not been able to promote previously.

To record used document shops

Many used record stores have signs out the the front. "We want your used data."

Consider seeking out data in bulk after which selling to used facts shops. Some people make a living doing this. Try to discover one store to deal with on a regular foundation and get to apprehend them. You will discover it profitable.

HOW TO SELL RECORDS TO USED RECORD SHOPS

1 – Set up an appointment with the document shop owner. Try not to make it a hectic tim together with on weekends

2 –Ensure that your information are in particular scenario. The record keep owners commonly search for incredible grades of information.

3- You can promote statistics in bunches or one at a time. Its absolutely as an lousy lot as you.

4 – Try first-rate file hold to promote to. I found that a few stores are keen for business agency and a few are not.

five –Ask the file store proprietor if they will deliver your inventory and pay you part of the sale at the equal time as they hold part of the profits.

Record Magazines

You have an great possibility to vicinity a classified advert in Record Magazines. The capability for earnings is proper because of the reality you are targeting record clients. It can be extra luxurious however its well worth attempting. Types of Record magazines

Record Collector

Spin Magazine

HOW TO SELL RECORDS IN RECORD MAGAZINES:

1- Studying the manner to growth an ad on a internet are attempting to find and make as attractive as viable. Stick to the statistics and try to draw humans to have a look at your gives. Used the phrases RARE, and GEM to your commercials

2 – Sell in bulk if you can and leave a internet site deal with for them to have a test. Leaving a internet site cope with also promotes your internet websites at the same time as offering the records for vsale as nicely.

DJs

Disc Jockeys don't use vinyl facts loads as they did because the vintage days. DJs depend more on computer structures to play songs. Look up DJs on the net and ask in the occasion that they have got information that they want to sell. A popular DJ also can have upwards of three,000 Vinyl statistics. A keep on Ebay began his business organization with inventory from a DJ and now he has a very a achievement enterprise corporation because of this.

Decide on a price to sell each record.

Please absorb consideration any transport prices as I reviewed in advance in the ebook. Don't cut your self quick. Plan for a 50 to 60% profits on every file your promote. If you're not effective look on-line to appearance what the vinyl document sells for. Put a sticky n be aware on each file noting the charge so that you are equipped while you positioned the file up to be had in the marketplace.

How a great deal to price

If you have got exquisite deliver of facts the temptation is to sell them off quick and make thousands of cash. But maintain on. You should likely supply away a pot of coins through manner of manner of wrongly pricing your report. This has occurred to me normally.

I as quickly as provided a Cramps document for $25. I regarded days in a while the net only to find out the equal report, equal circumstance, same pressing equal the whole lot presented for $ninety bucks.

Do your property paintings appearance online to appearance what the record sells for earlier than you promote it. Remember topics need to trade fast in this business enterprise. If there aren't too many copies flying round then the price goes up. If a band member dies or the band is folded the file ought to promote for appreciably extra money.

Likewise your price also can be too immoderate. Don't expect a not unusual multi dealer to promote masses higher than $10. Again research the artist test the condition and ensure you charge it fair at all times otherwise your shop will first-class have you ever ever looking at facts and no longer purchasing for them..

ITS ALL IN THE AD. YOU COULD HAVE THE BEST CONDITION RECORDS TO OFFER. IF YOU DON"T USE THE PROPER WORDS IN YOUR AD YOU'LL HAVE NO SALES.

When you charge your record soak up consideration the subsequent:

YOUR COSTS –On common:

- Cost you paid for the file

- Cost of a protector sleeve

- The fee of cardboard (if shipping)

    X 50%

  = Price to rate

- If you're selling online, your customer may be purchasing shipping.

If you're selling on Ebay and Amazon and your income are 2 hundred to three hundred in line with month that's a healthy aspect time income to have. You moreover must ask your self. "Do I make the equal cash imparting a mass quantity of records for sale and at a less pricey charge and achieving a smaller earnings. Or do I live with low portions excessive remarkable unusual steeply-priced gadgets. You want to check with this and be aware what works great for you.

Clean Your Records

Before you pass any similarly you want to go through every document and properly smooth them earlier than you promote them. Review the cleansing method I cover on this ebook. I'm very large on cleansing data. I assume this is critical in preserving an great saleable inventory

FOR THE RECORD

YOU COULD POTENTENTIALLY GIVE UP A LARGE PROFIT BY WRONGLY PRICING YOUR RECORD

How to easy facts:

Material required- Here is the fabric you'll want to smooth information:

• Lint Free Cloth

- Isopropyl Alcohol

- Water discipline

- Iron

Vinyl document should have a pleasing smooth shine to them. This manner no dirt, hair, finger prints, scratches. I have a look at my information under a strong light to make sure I see every flaw viable. Its hundreds higher you do this than you patron recognizing the problems for you . I recognize once I buy statistics off my Used Record business enterprise his data are flawlessly easy. He does a wonderful challenge thoroughly cleansing his Vinyl and this gives me first rate deal of self guarantee as quickly as I purchase from him.

You should purchase steeply-priced file cleaners that do an incredible task of cleansing vinyl however it's no longer vital to spend this money. All you need is a lint loose fabric that you could choose up from a dollar hold and a few Isopropal alcohol that you could choose up from a drug keep.

Mix the alcohol with water in a separate bottle. You want 70% water and 30% alcohol

combination. This interest is to make sure you don't harm the vinyl or the cover whilst you clean it. The alcohol diploma is truly enough to cast off finger smudges or dust.

NOTE: You can't remove scratches on data and maximum scuff. Once scratches are present on the floor of the record, they'll be there to stay. By cleaning your statistics you're making the overall look a good deal higher.

Vinyl cleaning – When you easy the vinyl try maintaining the file at its edges with the lint unfastened material. Try no longer to vicinity more finger prints on the perimeters of the file. Place a small amount of alcohol water fluid on the lint free fabric. One facet of the fabric want to be dry and the alternative aspect of the cloth want to be used for buffing and drying the report. Slowly dap the moist a part of the cloth across the report and wipe off proper now. If you allow it dry it'll leave a mark at the vinyl. Use the dry part of the material to softly buff out any finger prints or water marks left with the useful resource of the use of the wet a part of the fabric.

When you've wiped smooth each elements, look into it all over again under a sturdy lamp and location decrease decrease returned into the report sleeve at the same time as keeping immediately to the edge of the record.

Cover / Sleeve Cleaning- I've tousled many Record covers because of cleansing. Put a miles smaller quantity of alcohol water aggregate than you may for cleaning the vinyl part of the report. Your cause proper proper here may be to smooth off any finger prints or smudges.

CAUTION – If you positioned an excessive amount of answer on any part of the duvet, the card or paper may be completely broken.

If there may be pen mark on the duvet strive lightly rubbing it off with the solution mix. If it doesn't come off go away it and word it on your grading part of the document industrial.

Removing stickers – Sometimes you'll have the bizarre document that has rate stickers or large sale stickers on the duvet. Most of the time, I'll simply leave them on however some clients want no stickers on the covers. There are techniques to cast off stickers. One manner is to place a small drop of the alcohol water solution on the sticky label. Leave it in short at the sticky label. Don't allow it sink into the card cowl. Once the sticky label has enough moisture lightly remove the sticker from the quilt even as attempting now not to damage the cardboard.

You can also put off sticky label with an iron. Turn your iron on very low warmth and set the iron at the sticky label at the same time as constantly checking if you may pull off the sticky label. Once there is sufficient warmth finished, begin to peel the decal off. Be carefully now not to burn yourself! Also make sure that you're now not fading any hues on the record cover through making use of too much heat.

My report providers positioned a protecting sleeve on every record. This makes a wonderful difference while you're promoting facts. Having shielding sleeves on every file suggests your customers which you care about splendid. The shielding sleeves keep the report in accurate situation and enables reduce scuffing on the quilt. Look at the difference yourself. When you've got had been given a antique duplicate of a file and area a plastic sleeve on it, the scuffs and scratches seem considerably reduced. Your document will sell for plenty more through setting sleeves on record covers. You should purchase those covers from used report shops or on Ebay for a very low fee.

How to ship statistics

Don't make the error of delivery your facts thru manner of slipping them into an envelope and shipping them this way. There had been a few clients of mine that received report damaged. I smartened up and started out to put together a solid package. Remember facts are touchy and might destroy clean in cargo. The older facts from the early 60's and 70's are susceptible to breaking effortlessly as they may be made from a heavier plastic. Take be privy to this and make sure your cargo is well included.

Material you can need

- Scissors

- Clear Tape

- Cardboard

- Labels

Make effective that the file has at least ½" coverage extra around the sides of the package. I strong my package deal through setting a divider. I use this high-quality on older facts as they generally tend to break less complex. I additionally upload a very skinny foam sheet truely so no scuffing occurs during cargo. Try to characteristic a plastic sleeve to the file as nicely.

Watch that your package deal doesn't get to heavy. If you add an excessive amount of tape or cardboard you may get a wonder at located up administrative center when you find out how loads it's far to supply your package deal deal. Secure your bundle address multiple strips of tape on each aspect and corner of the package deal. Run a period of tape throughout the front and back of the package deal as protection coverage. Also buy an tremendous extraordinary easy packaging tape.

You should purchase report shippers from Ebay for approximately a dollar every. You don't need to try this. You can find out cardboard regionally at supermarkets or ask a nearby keep for a deliver. Just ensure the card has thickness sufficient to supply your facts.

A satisfactory clean package deal can tell one thousand terms. Don't make your package too sloppy. If the tape over hangs and the card isn't reduce at once you're sending a message to your client that you don't care and they may in no manner purchase off of you all over again. Keep it as neat as feasible commonly. I acquired a Record package from someone as quickly as. As fast because it arrived I knew that the record may be a multitude and it become. I need to inform just through the manner the report have end up packaged and shipped.

Labels - I commonly print labels through pc printer. It's neater and appears greater expert. Also I saved hundreds time this manner as an alternative then writing out the records on the cardboard cowl.

Shipping. - Don't deliver by way of way of ground mail till your customer asks you too. Ground mail is really too slow and the difference in rate is not well simply really worth the watch for your purchaser. If I deliver from Canada to US it commonly takes 7 business days. By floor it takes two times as lengthy.

Let your customer understand the shipping time constantly beneath promise and over supply. This will provoke your purchaser after they see how rapid their report has arrived. Shipping time is one the most essential capabilities on your sale. Remember your purchaser wishes the report in notable circumstance and shipped speedy.

Watch while transport file remote places. I discover the delivery time is exceptionally slow for some worldwide locations. Also hold an eye fixed fixed out for transport price. It receives as an opportunity high priced to deliver oversea. For example a report despatched to Japan may additionally need to cost as much as $24. Make you kingdom this inside the shipping fee. If you're no longer positive skip the positioned up office and ask for an estimate first.

It's quality to get a tracking huge range together collectively with your document transport. Your purchaser also can need to apprehend in which their package is and its proper to have this facts prepared.

If you're selling via mail order possibly one in every of maximum crucial matters to observe is how you could get your report there as it have to be and on time.

For the quilt

A lot of my customers buy the record most effective for the duvet. They just like the reality that they will dangle up the quilt like a portrait in a frame which you actually can't do with a CD. There are a few recorc covers which may be well well worth masses of cash so long as they're in accurate state of affairs. There are some humans which have offered document from me definitely to get a hld of the cover. The older 50's and 60s' covers are solid more inflexible cardboard and feature the artist picture observed on a separate glued photograph.

 Some of my clients have even requested no longer to supply the record. They simply are interested in the duvet. In reality I observed shop and flea markets in reality selling covers of the records.

When you're selling your file, nation the condition of the quilt because of the reality this is an incredible promoting characteristic.

- Where to find vinyl file.

Flea Markets – We have big flea markets near our domestic which have lots of vinyl facts. All of them are in Mint or Near Mint situation. The average rate is generally $four to $5. If you in bulk they typically take down the price even greater. This is in which I purchase I majority of my Vinyl. I typically purchase 10 to fifteen statistics at one time.

DJS- QUITE A FEW DJs ARE NOT USING VINYL RECORDS ANY LONGER. Most of them have transformed to laptop structures to play song. Place an advert in a song mag or nearby paper and ask if any DJs have an antique stash of facts to remove.

On line -Ebay – I certainly have presented quite a few records off Ebay in modern years. Ebay is so clean to promote and buy on. The quality way to shop for on Ebay is to buy in bulk. Try to discover a reliable provider. Find one which owns a store and artwork up a address them. Try to discover nearby Ebay stores. This will maintain your delivery fees lower. Look on Ebay for Wholesale masses for starters.

On line - Music Stack – This is in which I get all my facts from. There are over 1 million information to be had on Music Stack with hundreds of various suppliers. I discover Music Stack to have the maximum critical selection of information as well. I deal with companies who can fill all my client's needs. If they may be capable of't come through for me there are plenty greater of sellers which could help me out.

On Line -Gemm Records – I don't recognize a whole lot approximately this internet site on line. I haven't use them however . They also deliver a big record collection from more than one suppliers. From what I've seen they appear they've got a extraordinary comments and fee system. Try them out and have a look at what you think.

On Line - Private websites – You can surf the Web and in reality discover loads of net page that sell Vinyl statistics. Try to hook up with a few them and spot how they may meet you purchaser dreams.

Locally – Place an advert in your close by paper and try and run it for as long as you could. You is probably surprised what gem stones you may find out. There might be someone geared up to sell off masses of records. Just ensure you inquire about the high-quality of each record.

Garage Sales -Don't stay away from Garage earnings. Alot of humans do spring cleansing or are planning to transport. They ought to decide they don't want data anymore. If you spot a storage sale taking place zero in and search for records that they will be promoting. Most of the time you can discover facts dirt reasonably-priced.

OR Don't purchase your stock – Keep your stock collectively collectively together with your companies. I run my employer through drop transport. I actually have agencies that supply the stock. They p.C., supply and I deal with rest of the transaction.

HOW DO ARRANGE FOR DROP SHIIPING OF YOUR RECORDS

•   Search on Music Stack. It's the extremely good place to discover a drop shipper.

•   Once you discover a shipper ask them if they may supply data to your customer. Ask them for a wholesale fee. Also inquire approximately amount reductions.

•   Order some statistics and function them sent to your private home first. By doing this, you could audit the shipping time and high-quality of the data.

Chapter 5: Take A Listen

Take a be aware of each document before you sell it. You can also use this as a awesome selling function. If your patron is aware about which you pay attention to every record that gives them searching out self notion and maintains them coming once more for additonal.

It's now not hard to do that. My record hold supplier listens to every report in stock. He performs the data as he works. By being attentive to report it additionally gives you more information about the file so that you can answer any questions about the LP or unmarried.

Skipping – Take phrase on the identical time because the file skips. Make be aware of the songs that pass and encompass this data to your report advert. There's not an lousy lot you can do to correct this aside from informing your functionality purchaser of the flaw.

Static – Static can be due to a number of of things. It may want to happen whilst the plastic sleeve rubs in opposition to the vinyl as you're getting rid of the disc. This generally takes vicinity on more modern day pressing of Vinyl Records. The fabric used is lots thinner and extra widespread. You can usually cast off this with the aid of the usage of the use of strolling a rubber Vinyl roller over the document. It can choose any ultimate static that's left at the document whilst you dispose of it from the sleeve.

Scratches- Scratches are in reality as demanding as skips. By looking on the vinyl underneath strong mild you may usually see scratches on Vinyl. Not all scratches are giant inside the course of the gambling of the report. Include this facts to your document advert. Let the client apprehend that there are scratches visible however they don't motive any issues at the same time as gambling the record. You can try to buff out scratches on data with you lint unfastened cloth however there wont be masses of a difference in how the document appears.

Scuffs – Scuffs are due to the internal sleeve rubbing in the direction of vinyl ultimately of shipping or motion of your package deal deal. Some of my customers inquire from me to deliver records out of doors of the file sleeve in order that the vinyl doesn't rub towards the plastic sleeve. You can reduce the quantity of scuffing a document receives with the resource of restricting the movement of your facts and packing it tight at the same time as you ship it. Try to line up statistics them up tight internal bins. You'll word a record maintain want to hold the statistics tight in the shelves in order that they don't bypass round lots. Scuff marks may be reduced with the aid of the use of way of buffing out with the lint free cloth.

Take photographs of every of your facts whether or not or no longer you're going to sell to report shops, flea markets or on line. Remember taking superb images may be very crucial to sell your facts.

How to take images.

A accurate photograph allow you to make the sale. Take some time and take a pleasing shot of every report that you have in stock. If feasible eliminate the vinyl disc from its cover and take a photo of the cover and the vinyl on the identical time. Try to seize the label on the disc and the decision on the cover.

Taking a clean picture showcases your data on your capability customers. It captures the best of the record at the same time. When your patron sees the pix it moreover suggests them which you're prepared and feature well product to promote.

If feasible try and dispose of the plastic defensive preceding to taking snap shots of your document. Occasionally the flash may additionally additionally furthermore purpose the photograph to emerge as distorted with a large white flash interfering with the photo and making it hard to appearance the quilt. Try to use a pleasant background for the file. Don't use a colour historical beyond a good way to combination toward the shade of the quilt.

Keep your snap shots to your laptop documents. If you sell a file you can update it with some other one of the equal pressing if feasible. In this example if the grading is the equal you have no want to retake a picture.

Many of my clients ask approximately Matrix codes or etching on the indoors part of the ring document vinyl. A majority of vinyl records have this. When you check file etching as its known as, it almost looks as if someone's handwriting at the disc. I however haven't discovered out how that is pressed at once to the record. For some cause the greater moderen vinyl record don't have those codes etched at the inner ring. Believe it or not a few vinyl facts have humorous comments on this a part of the report, no matter the reality that I actually have not all started to peer one.

These codes have to imply one of the following topics:

- Record Label Catalog or LP Number

- Number Mastering Codes

- Pressing plant of the document

Sometimes certain codes pick out the document as very rare and expensive. Take a examine your records and spot in case you be conscious these codes as you're cleaning your information.

There are also codes on the label of the file as well. The code at the file label in shape the code that at the the front or back of the report cowl. These are essential codes which you need to encompass to your advertisements. They should constitute the call of the document organization and pressing data. Some artist produce multiple pressings of the same file. When you have a look at the codes on the duvet or the label it have to provide you with most of this information. More information desired

FOR THE RECORD

If the report has a EAN code on the quilt it approach it turn out to be made after 1973

VINYL AS ART

Occasionally you may have vinyl facts for your collection that aren't playable and also you may think they are not saleable. Don't soar to brief assumptions on this. You can though buy loads of unplayable vinyl facts and sell them as art work portions.

Many of my customers purchase the record only for the cover. They similar to the reality that they will draw near up the quilt like a portrait in a body that you really can't do with a CD. There are a few report covers which might be well worth plenty of coins as long as they will be in well condition. People have bought data from me just to get a preserve of the cover. The older 50's and 60s' covers are stable greater inflexible cardboard and characteristic the artist image posted on a separate glued photo.

Some of my customers have even requested not to ship the record. They actually are interested by the duvet. In fact I found preserve and flea markets definitely selling covers of the statistics. Another dealer I understand promote file cover and make clocks out them

When you're selling your record, country the circumstance of the quilt due to the truth this is an notable promoting feature. Your client may moreover really need the quilt

Selling Your records as Art?

1-   Some artists soften antique facts into ashrays or pencil holders. They make sure the labels are no matter the truth that visible so that you can see the artist's name.

2- There are some collectors that wax the vinyl into a very shinny state and region the used vinyl report in a body with the artist's photo as a tribute to the artist. Many people have computer systems at home so upload pix to the Vinyl disc and increase incredible awesome images

three- Some accumulate the covers, easy then and region them in frames with lovely velvet backgrounds.

four- I've visible a few creditors acquire the labels through putting aside them from the vinyl and setting the labels in frames as a tribute to the artist.

So don't throw out or offer away your document collections that aren't playable. There is continuously someone obtainable that could want to shop for them for artwork.

• SETTING UP A RECORD STORE.

I can tell you putting in on-line Record Store may be overwhelming but the rewards are well properly worth it. It's in reality no longer any much less difficult than putting in a brick and mortar save. You need to truly focus on doing a terrific process however at the same time as you're putting it up. The greater try and care you located up the the front the better the very last results of achievement. There's opposition obtainable but don't permit that scare you. There's masses hobby in vinyl information and you have to examine this opportunity to make some of cash.

Whether you've got got had been given a flea marketplace place, brick and mortar keep or on-line store make your preserve attractive. Ask your self, what's going to preserve my customers coming once more. Quality, velocity easiness to sale, and treating customers like gold are key proper right here.

Offer shipping reductions, unique information on sale, quantity reductions as example to bringing clients once more.

Make your keep outstanding and cheery. Use outstanding excellent hues for your commercials or historical past.

Before stepping into installing net net web sites, use Ebay for starters. Ebay has a following of 18 million human beings. Work on this for a few weeks than begin increasing from there. I can sincerely let you know in case you're strolling more than one shops at one time, you're certain to make an superb residing at this. Each and regular your competencies will boom with every sale and you'll end up an expert.

Chapter 6: The More Stores The Better

Start Ebay with a $15 a month store and use your income to open at $40 nine a month save. When you open a $15 a month keep start with 2 hundred items of blended accurate outstanding records. Build up some accurate remarks at the same time as you start. Feedback is given via your customers. Its a system that quotes the incredible of your company Feedback is essential as your consumer may want to have a look at this and decide whether they can buy facts from you. You may additionally moreover want to begin purchasing for a few information first and begin constructing your remarks this manner. Register with Ebay first. Its very smooth and on-line help is usually to be had.

Then take a look at in with EBID. They are quite low charge to join up with and function a following of about 2 million human beings. I advise you to set up a store right here as properly. You can location as many commercials as you need right here for a totally low price. I propose you operate different people inventory and list hundreds or heaps if you could.

Then join up with Ecrater. They are free and developing unexpectedly. Place as many record profits adS proper right here as viable. Be high-quality to open a shop.

AFTER SETTING UP STORES ON EBAY, EBID AND ECRATER SET UP A FREE WEB SITE AND PROMOTE ALL YOUR RECORD SELLING THROUGH GOOGLE AND YAHOO MARKETING TOOLS.

If you start with those 3 internet websites you'll take a look at you'll sale developing weekly. Again don't fear about stock. There masses a locations to get your deliver of facts as I stated formerly. Be affected man or woman with on foot shops. Its all about recognition. If you can develop desirable don't forget at the side of your clients your document profits will stay sturdy all the time.

The first-rate problem about installing on line file shops is whilst you set them up with thousands of information to sell, a drop transport technique its outstanding easy to run. In truth I love how easy it's miles making income as you sleep. Good success with this!

• RECORD THICKNESS

Back within the vintage days, facts made within the 60's were constituted of a far heavier grade of plastic that the extra moderen information made inside the seventies and eighties. It wasn't until the seventies that document agencies started the usage of a lighter weight flexible vinyl to reduce fees. Other businesses then started to press records on recycled vinyl appeared Dynaflex 125G vinyl. The greater moderen recycled vinyl have masses more vast pops and ticks because of the reused plastic. If you examine the vinyl facts made within the specific earlier a long term you'll look at the greater moderen the vinyl the an lousy lot extra flexible and it seems tons better made. In fact you can bend a contemporary document and if attempt doing this with an in advance file you most probable will damage it. One problem of newly made facts is that if you don't save or supply them well they may warp.

180G Older Vinyl –   Advantage – Sound better

                                  May   remaining longer

                                  Looks   better   to lenders eyes

                                  Higher   name   for from lenders

                                  Disadvantage   –May Break in shipping or garage

                                  Costlier to deliver

More steeply-priced to shop for

125G Newer Vinyl – Advantage – Costs a high-quality deal a whole lot much less than older thicker vinyl

Cheaper to deliver

Easier to discover

Disadvantage – May warp in transport or garage

More ticks and pops evident at the identical time as playing

When I first began out out some of my clients who acquired records via mail should complain that the file changed into broken despite the truth that I took each precaution to ensure it changed into packed efficiently. It doesn't take plenty for an older report to crack so make sure you % it with more care.

Something of Vinyl thickness you have to recognise approximately while promoting data

Record thickness moreover subjects to creditors. Some lenders are requesting one hundred eighty gram vinyl this is the heavier thicker vinyl. Collectors claim the180gram vinyl weight has a far higher sound pleasant than the thinner normal vinyl information. Apparently a few lenders say the thicker vinyl lasts a exceptional deal longer than thinner vinyl and you can get loads more playing times that the ordinary vinyl. I count on that may be a fantasy. I see no purpose why the thinner vinyl can't have the equal amount of plays as an older thicker vinyl record. The heavy vinyl is likewise known as Virgin Vinyl or 180gram thickness vinyl. This form of vinyl changed into utilized in older pressings and additional trendy pressings of classical track.

EQUIPMENT YOU NEED TO START SELLING VINYL RECORDS

1.   Another great benefit to selling information is that you don't want to shop for a number of tool. In truth the principle piece of device you need is a pc and that's it. This is particularly real if you plan to maintain minimal stock.

2.   The 2nd most important detail you need is a record player. It's crucial that you have a reliable piece of device proper right here. You can spend as low as $100 or as plenty as $three,000 proper here. Shop spherical however make sure you buy a splendid brilliant turntable to check your facts. Do your studies in advance than you purchase. You may additionally moreover contact some DJ's. They don't use turntables as a bargain in their track is now brought thru the laptop. You'll discover you may be capable of get some relevant gives proper right here.

three.   You additionally need to buy bunch of sturdy containers to store your information. Make powerful they will be of strong higher first-class plastic. This is simply so they don't fall apart at the same time as you stack then one on pinnacle of each different. Go to Walmart for the exceptional gives on plastic packing containers.

4. Buy companies for cleansing. Make fantastic you buy Isopropal alcohol for cleaning. Buy some lint unfastened cloths as well. You can choose cloths up at your community dollar save.

five. Plastic sleeves. Go you Ebay or ask a used report keep to offer you with plastic sleeves to protect your records in shipment or storage.

Below is a file promoting assignment list which I recommend you to use. Review the entirety in this list and make sure you have got got had been given the whole thing in order in advance than you begin selling data. Also upload something you need that you could take into account.

TASK COMPLETE

Do I certainly have sufficient inventory to start promoting data - (minimum two hundred)

Do I actually have an outstanding choice of facts

Are my records wiped clean

Are my statistics easy to find out and taken care of

Are my data stored in a suitable dry easy room

Have researched to look what my information can promote for

Am I persevering with to search for inventory to feature

Did I begin to prepare some advertisements for my facts to sell

Have I three to four unique way of promoting my facts

Have I signed up with online internet site online a very good manner to get my stock listed

Have I appeared into renting some flea marketplace location

Have I requested about renting some ground place in a ebook store

Do my expenses deliver my an normal 60 % earnings in selling

Have I set goals for my document promoting company

Did I look on how one-of-a-type record busineeeses are doing on line

Have I seemed everywhere to locate invenotry for my information

Do I understand sufficient approximately the document promoting commercial company to start promoting

Have I taken clean pictures of all my inventory

Have I concentrate to my file stock

Have I graded my statistics steady with the cutting-edge condition

Do I sincerely have the right gadget to begin selling records.

Have you in assessment other report net web sites stores

Have you tested to deliver a report inorder to appearance in case your packaging is enough

Do you have were given sufficient packing materials

Develop some Ads

You want this whether or now not you're selling on-line in a newspaper or at a flea marketplace. This is one of the maximum crucial steps to selling your data. Refer to the segment on growing a ad within the e-book There's three varieties of ad I communicate approximately. Make positive you provide sufficient records about your records. The type, situation and whats unique approximately it.

• How to installation a report selling advert. Your ad will each kill your sale or supply it to earnings. I'll assessment what it takes to get a sale.

Here I'll display you the manner to well set up a record promoting ad. The manner you lay out your advert will each kill or ship it to the check out basket. You need to understand and be conscious what it takes to make the sale.

I've tried it each strategies. I've said very little about the document giving simplest giving a few detail and moreover attempted over doing it and giving quite a few information. I positioned that when I gave information approximately the file it sold an awful lot better than giving little element.

I'd furthermore like to add one problem make your information accurate as nicely. The average document collector may additionally additionally seize on incorrect records you've supplied after which they obtained't come decrease returned on your save.

When you offer statistics for your advert encompass the following

Artist Name

Record Name

Country of Orgin

Type of Music

List of Songs

Size of Record

Grade of the Record

Chapter 7: Shipping Time If By Means Of Using Mail

Bonus facts that sincerely works – brief tale approximately the document.

I've positioned that offering a brief tale on the report significantly will boom your opportunities to promote information. It moreover offers the patron the idea which you realise your stuff and that you care approximately the sale. Many of my on-line clients have said how impressed they'll be with the facts I've given them at the record. Where do I get this records? Usually from the net or without a doubt primary expertise I without a doubt have about the artist.

But this exceptional selling advert tool shouldn't be used most effective for on-line earnings. I've attached this records to statistics I've bought at flea markets and in used document stores and it works exquisite. Look on the photograph underneath to look what I advocate. Look at the seen distinction with a record that has no statistics related to a report that has a sleeve and the proper statistics linked. It's a massive distinction!

Here's an example of the ad for the facts In promote on-line. More infor needed

• Some of my unique document selling evaluations.Its types of crucial that I proportion some of my studies correct and horrific so you can do higher.

I've been promoting statistics for years and I without a doubt have a few high-quality recollections to tell. Most are remarkable reviews and I choice some of these reports assist you to come to be successful to promote data.

Buying –

When I decided I preferred to make coins selling data I desired more a selection and amount. I located this flea marketplace just an hour north of me that have   location which awesome remarkable facts at a top notch charge. I commenced out buying from this vicinity and this supported my initial start at the business enterprise. Why I'm telling that is that some people consisting of myself assumed that data offered at flea markets are low amount low top notch but in a few times this isn't genuine. Try shopping for spherical at flea markets and also you is probably amazed what you'll find out.

Selling

I began out to list records on-line which I didn't have. I knew the situation and cover

and the entirety else approximately the report however I didn't very very very own the document. My pal on the neighborhood document save stored I statistics I wanted in his inventory. When the file bought I went to the shop and bought the record off of him and shipped it to my customer. This stored me large inventory charges as someone else come to be keeping stock for me. I advanced a clearly precise rapport with this file store. You might likely ak "well what about if he bought the stock" That's smooth. I maintain checking his website on line to look his up to date document stock. If I decided through manner of risk the file was offered, I typically obtained one from some other place. It become that easy.

Next Phase of Selling – Oh What A Feeling

I persevered to listing records which I didn't have. This time I positioned providers on Music Stack and started out out list the loads of records that have been to be had at this website. Once I made the sale I despatched them the information and that they shipped to my consumer no questions asked. They billed me and I paid them via Pay Pal. My income genuinely took off .I had lots extra fashion of records that I may want to make to be had to my clients. I moreover advanced a real self assurance of the financial enterprise at this issue. I wager at this component I knew what I turn out to be doing and I felt I need to make well coins promoting Vinyl statistics.

My Biggest Order

My biggest order turn out to be from someone in Russia favored me to deliver him 100 and seventy facts. Everything went properly till he declined to pay the delivery charges Thecosts were very immoderate however he changed into getting a bargain on the information. Too awful although it'd were a nice earnings!

Insane Customer

I normally reply to client questions right away. There turn out to be one time wherein a capability consumer asked one query then each different then some other. In all I replied approximately 15 questions on this one record. Finally I said to him "Look you don't want this document do you?" Even if he desired the document I assume I wouldn't promote it to him in any case. He truly didn't look like someone that I'd want to cope with.

Most disillusioned Customer

A customer of mine in New York ordered one of those 180g 1960's thick vinyl file with the beneficial aid of Womb. It have end up in top notch state of affairs while it left however a few days later he despatched me an unsightly email pronouncing he become dissatisfied that the file broke for the duration of shipment. I gave him his cash over again however he stated it didn't make distinction because of the reality he really desired this document. He ended up filling a grievance on Ebay approximately me despite the fact that I advised him how sorry I became. Some collectors are excessive about their records. When they make a incredible find out they need it in best situation and shipped fast. From that day on, I determined out my lesson and I deliver with none troubles.

Most Coolest Customer

I bought a file from a close-by flea market. It changed into absolutely scuffed up on the out of doors but the vinly became extremely good and easy. It as a 12" Dead or Alive unmarried. I wiped easy it up positioned a plastic sleeve on it, informed a touch tale about the record provided it at 3 instances the price I supplied it at. I come to be virtually involved that who ever sold this could nevertheless bitch about the situation of the record. I sold it to someone in southern US. Boy modified into he glad. He informed me how a splendid deal he desired how I packed the document and the manner exceptional the story I told about the file become. He gave exquisite comments and characteristic emerge as a ordinary client of mine. Sometimes it doesn't matter range huge range approximately if there are a few minor flaws to your facts which you sell. As prolonged as you treat your cuastomers like gold and do your amazing to provide them something this is suitable than you're earlier of the sport.

Shipping Story

I shipped a file to Japan as quickly as and made the error of transport by using ground. Actually it wasn't absolutely my mistake. My customer asked it shipped this way. I don't assume he realized or I located out that it'd take over a month to get there. He saved sending me electronic mail after e mail. I in the end sent him his cash returned. I in no manner instructed him it have become his fault for ordering it this way. I in reality refunded him. No questions requested. When he had been given the file he despatched me a nice message and refunded the cash lower again to me. He said I actually have end up sincere and I virtually concept he modified into for being so sincere sufficient to pay me for the file.

I'm sure you can in the end have hundreds of stories inside the destiny to percentage.

Learn form your customers and adjust your Record Selling industrial enterprise in the end based completely on the remarks you get out of your profits. Good or terrible.

• Where to get records at the fee of your information. I'll evaluate in which to get information at the rate of your information and its continuously the net.

As you benefit increasingly experience selling your Vinyl statistics you'll quite hundreds recognise how a bargain a document will promote for. For example I apprehend that led Zeppelin IV doesn't promote less that $20 for a Near Mint copy.

This is a totally grey vicinity for some because there can be a lot competition to be had an entire lot of misunderstanding is going on the aspect of it as well. Many file dealers below cut their rate of the Vinyl first-rate to find that they lose cash in the long run.

There are two strategies to cope with this:

1- Through experience selling

2- Look on line to discover what the file sells for

3- Record appraising magazines.

Chapter 8: For The Record

Don't beneath lessen your rate simply to promote your report. You'll lose profits and a possible an extraordinary report. Place a nice advert, wait and someone will sooner or later purchase it.

Here are Record Appraisers I use:

The Record Finder is a famous internet site which I use. It has a hundred,000 file catalog so that you're maximum positive to discover the charge of your facts proper right here. They additionally consist of fees of every report indexed

Vintage Vinyl is likewise a super vicinity to do studies on Vinyl record price. There have a fantastic series of statistics as nicely to browse through.

If you need to promote funk or Jazz Vinyl LP's then you definately definately must checkout Dusty Groove America. They are offer you with an correct charge of your Vinyl

The Princeton file alternate is able to provide you with an close to concept at the rate of your Vinyl as well.

I moreover attempted Google. When you would like to appearance up a report, kind in the name using Quotations and ensure you kind within the phrase vinyl at the cease. You're first-rate to offer you with some thing right right here.

I additionally take a look at record auctions which may be taking location Ebay. Don't typically rely on this in spite of the truth that. There is a lot of competition in this website and from time to time I truely can't inform the charge as there can be a lot difference in fees.

Lastly I usually walk thru used vinyl records stores and take notes of what prices they may be selling their data for. I discover that I can observe the situation on the same time. This way I can in form up the grade to the fee of the file.

- Keeping your stock organized

Keep Good Records

Make incredible what you've offered and As I've said in advance than replace a document that you soolld as soon as feasible. If you've sold it as quickly as you can sell it another time and maximum likely at a better fee. Keep easy documents on what you purchased, who you obtain it to, even as you got it and state of affairs of file.

There's not some thing more worrying than getting an order from a purchaser and now not being able to discover the file. You'll will find out while you start to deliver masses or loads of Vinyl records it becomes pinnacle precedence to be prepared specifically even as storing statistics in stock.

I typically hold my record prepared alphabetically. It's the correct maximum secure approach of doing so. By doing this I know precisely in which to discover my statistics. I list them alphabetically with the useful resource of artist first.

Some sellers I 've seen in file shops prepare their vinyl by using manner of way of genres of tune. This commonly works well with some that has a big quantity of facts. They can also need to listing in this order music then listing the artist alphabetically.

Believe or now not I've moreover visible record lenders put together their song with the beneficial useful resource of release date. I'm now not certain how well this works but for dealers that need to recognize how many antique facts they have got they opt to list in this order.

Many dealers put together their statistics based totally on rarity. They preserve the tons much less valued information separated from the better priced coins makers. I can see the common feel in this. I someday experience that I want to maintain my unusual statistics in a stable secluded spot until I'm geared up to promote them off. I in reality don't like them mixed in the ordinary populace of Vinyl.

The super development for you as quickly as you've got were given a giant amount of information in stock is download a unfastened facts base software from the net and installation your records on this record. When you promote a record visit your statistics base and replace the documents. Set it up so it tells you on the same time as re order a record that has been offered to your inventory.

More statistics desired on statistics bases and so forth...

• What do record grades advise and the way they may be capable of have an impact for your earnings.

People will buy your inventory of records if you kingdom the grade of the record. The grade manner the contemporary-day situation of the report. It additionally manner a sincere grading of the report.

Its no need giving the purchaser a faux grade because of the truth the customer will find out the truth and in no manner come again.

I use a grading tool for all my information whether or no longer I'm promoting online or at a flea marketplace. Its one of the first question a purchaser will ask. Just to will will let you apprehend severe collectots are stern regarding the accuracy of the grade. If you preserve on with proper grading statistics in your statistics you could't cross wrong.

I use The Goldmine Grading as it's far the maximum diagnosed tool for grading statistics.:

On a scale from 1 (Poor) to ten (Perfect) the above gradings are same to:

MINT - 10, Near Mint - eight, Excellent - 7, Very Good Plus - 6 , Very Good - 5 , Good - 2

PERFECT — There isn't this form of component as a PERFECT used report. If a person tells you the used record is first-class they will be not being sincere approximately the grade.

MINT (M): This way the Vinyl Record is right. It generally manner it simply left the presses and has no flaws the least bit. The vinyl has no scratches or scuffs. It seems one hundred% smooth and shinny. The cowl has no stickers on the the the front or once more. The cover is not worn in any respect. If you've got have been given record together with this, congratulations. It doesn't advocate that the file must be new. There are some antique data which may be in mint state of affairs. Be cautious at the same time as you assert mint it is mint. Most of my records are near mint. I really can't find out a mint report except its pretty new. I'm trying to be as honest as viable.

NEAR MINT or NM, M- : To be close to mint the record has to have very minor scuffs and now not masses else to gather this grading. It may additionally have very minor marks on the duvet however that's it. The cover want to examine although its in best condition. Whatever marks that the record has on the vinyl or the cover should be very minor. When you ply the file it shouldn't have any noise in any respect. As I stated formerly maximum of my records are in Near Mint situation. This is what I look for once I pick out a record to buy.

EXCELLENT or EX or VG++ : A Excellent record should visible scuffs however they need to be very minor. Compared to Near Mint you ought to be capable of see them with out searching too tough. No seam cut up on the covers The vinyl want to but comprise no scratches.

VERY GOOD PLUS or VG+ This file might show some put on and ground scuffs. The flaws on the vinyl are nevertheless seen but the vinyl however seems smooth. Still there need to be no floor noise the least bit. The cowl may additionally moreover additionally have a few writing on it however apart from that it must be very clean to make this grade. The corners of the covers may also have moderate canine ears but don't have any tears inside the backbone or holes. As a rule there have to be no more than three flaws to make this grade.

VERY GOOD or VG: There can be a number of flaws with this record however the file stays appropriate. The ground noise is gift on the same time as the file plays. It may be obvious at some point of the file. If the vinyl appears clean it ought to even though meet this grade.. The cover could have a few placed on to it along facet some ring put on. It will appeared elderly in evaluation to as more moderen report.

GOOD or G A proper file need to nevertheless play nicely. There is probably some crackling noise within the direction of the document.. The vinyl condition will appearance as although it's been completed frequently and you can visibly be aware the wear and tear and tear. There want to nevertheless be no skips. You need to nevertheless be capable of play the file and experience it at the same time. The cowl will look like it has some large located directly to it. It can be taped on the seams and writing in markers on the the front or all over again.

POOR –Usually is reserved for a report graveyard. This is wherein the document can be recycled in to an ashtray or pencil holder. The cowl if it's proper may be used for paintings.

- How to sell a ton of statistics. Treat your customers like gold. Offer 100% coins lower again guarentees.

Step 7 – Review – Look at your statistics. Do you feel you have were given a brilliant desire? Good great, wiped clean and a tagged at a first rate rate? Are you prepared to begin promoting on line or regionally to report shops or flea markets. Are your information organized so that you or anyone can locate a few component inside seconds?

When I go to my pal who owns a file store and I ask him for a report he usually well-knownshows something internal seconds of me asking.

Make excellent the entirety seems correct in conjunction with your inventory in advance than you start promoting.

I attempted numerous techniques of selling vinyl information and I need to percentage what works for me:

Treat your customers like gold. Make them apprehend which you'll do some thing is important to maintain them attempting to find shape you. My buddy at the used file keep knowledgeable me he appreciates my commercial organization. Whenever I ask him to get an incredible report for me for me. He goes out of his manner to get me what I need. You have to do the equal issue. I constantly will pass back to him because of this.

Try to provide a wonderful priced item. Don't over fee. Sell at a cost in which you are making some income at the identical time as on the same time you're providing a exquisite deal for your consumer

Offer coins back ensures. Remember on the same time as you promote loads of statistics there might be one or customers that gained't be glad with their product. You want to provide them their cash lower back irrespective of what. If the record is uncommon and precious to you're making arrangements to have the document sent yet again you. Otherwise have then maintain the document.

Have a smooth splendid product. I truely recognition on this. Make sure the report is smooth and in appropriate circumstance otherwise don't promote it for a excessive charge. Ensure it has a plastic sleeve, has all of the statistics and a tale connected to the quilt.

Sell throughout. The internet flea markets, newspaper, loose websites, pay net sites, stores. Try promoting everywhere. Don't use first-class one outlet. The world is yours to sell you vinyl information. The greater manner or shops you promote to the more facts in line with month you'll promote.

Have a whole ad. Give as lots records you may about the document. Be honest, be honest approximately the price within the ad. Don't cheat. .

Have a laugh with it. I've had a lot entertainment promoting data. I love telling my buddy Im inside the song commercial enterprise company and that I make appropriate cash doing it. It also very clean. There's hundreds of labor up the the front however then you definately coast and everything starts rolling.

Lets in quick over the stairs to begin promoting Vinyl Records – Reality Check!

Step 1- Gather your stock

Your inventory is each made from facts you've had mendacity round or they will be items your purchases. Hopefully you have got an inventory of at least two hundred to 500 minimum to begin. The first difficulty you need to is kind your facts. I typically kind them alphabetically for a begin. Then I label them. I'll speak further in this a hint later I the ebook.

Before you bypass any in addition you want to research every record and deliver it a grade. The cause why I say this in advance than cleaning is that cleansing obtained't get out all scratches and scuffs at the file. Use the grading gadget which I noted within the ebook. Put tag (show photograph)

On each report:

Name:

Format – LP , Single or 12" Single

Grade: Mint, Near Mint, Very Good, Fair or not saleable – Cover saleable?

Step 2 – Clean Your Records

Before you pass any similarly you should undergo each report and nicely clean them earlier than you promote them. Review the cleansing approach I cowl on this ebook. I'm very big on cleaning facts. I assume this is critical in maintaining an extremely good saleable stock

The first cleansing is critical earlier than you located your information away in inventory. You'll clean your statistics all over again earlier than you entire a sale. A 2nd cleansing is needed as your data may be sitting gathering dirt if stored for longer intervals of time.

Step 3 – Decide on a price to promote every report.

 Please soak up consideration any shipping prices as I reviewed earlier in the e book. Don't lessen yourself brief. Plan for a 50 to 60% income on every record your sell. If you're now not positive look on line to peer what the vinyl file sells for. Put a sticky n word on every document noting the rate so that you are equipped even as you put the record up available on the market.

Step 4 – Develop some Ads

You want this whether or now not you're selling on line in a newspaper or at a flea market. This is one of the most important steps to selling your statistics. Refer to the segment on developing a advert inside the ebook There's three forms of advert I talk approximately. Make certain you offer sufficient records about your data. The kind, situation and whats specific about it.

Step 5 – Decide Where to Sell

If your going to promote to used Vinyl shops installation a time for the owner to vicinity apart which will test each certainly one of your statistics. Try no longer to sell your entire collection without delay. Attempt to connect with multiple supplier to promote and buy records from. If it's Ebay or Amazon as an example test in with each the sort of web sites in an effort to begin listing your statistics.

Step 6 – Take photographs – Take photos of every of your statistics whether or not you're going to sell to report shops, flea markets or on line. Remember taking suitable images might be very critical to promote your records.

Step 7 – Review – Look at your information. Do you feel you've got a exceptional choice? Good remarkable, wiped clean and a tagged at a first rate rate? Are you organized to start selling on line or locally to file stores or flea markets. Are your facts organized so that you or everyone can find out some element inside seconds?

When I go to my pal who owns a document keep and I ask him for a report he usually exhibits something inner seconds of me asking.

Make positive the whole thing seems correct collectively together with your inventory earlier than you begin selling.

Step 8

Start Listing

There are many regions you may list your Records – On line, Used Record shops, Flea Markets or your very personal non-public Web internet website online.

On Line – Ebay for example – Start small and open a $15 a month preserve and start listing your merchandise. Stay far from auctions. I don't have any patience for this. I'd as an alternative list a record for one month at a time

Used Record Stores – Set up an appointment with a used file supplier. Try to growth a incredible relationship so you also can purchase from the proprietor as well. It's very critical you maintain you facts prepared and easy as they'll pay you more money in your facts.

Flea Markets – Try to rent a small desk at you community flea market. Flea Markets are first rate for profits.

Note: If you could sell at three region indexed above you're positive for tremendous fulfillment!

Chapter 9: The Vinyl Revival

Vinyl or Digital. Both have their advantages and downsides. Audiophiles might claim the sound superb of vinyl is advanced, others may argue that notion. But in case you are studying this then you definately definately possibly revel in that vinyl is the most accurate and emotional approach of sound reproduction. And right right here's why.

Vinyl facts create frequency, amplitude, and wavelength by using way of the usage of a bodily grooved ground to record sound, ensuing in a more precise and accurate duplicate of the proper audio recording. The physicality of vinyl records and the analogue nature of their sound create a totally specific listening experience that many audiophiles discover extra sensible and real than virtual options.

Why could possibly that be? Because vinyl information offer a heat and herbal sound that isn't always confined with the aid of digital choice, making them the medium of preference for audiophiles and tune enthusiasts who understand the best features of analogue sound reproduction.

The Science of Sound

To better understand the relationship a few of the not unusual-or-garden file groove and the stylus, it's beneficial to go once more to the fundamentals of tactics sound works.

How the human ear perceives sound

Sound is a shape of energy that travels through the air inside the form of waves. When sound waves gain the human ear, they reason the eardrum to vibrate, which in flip devices off a chain of activities that permits us to recognize sound.

The human ear is made from three essential elements: the outer ear, the middle ear, and the inner ear. The outer ear is the visible a part of the ear, which funnels sound waves into the ear canal. The center ear consists of the eardrum, and 3 tiny bones referred to as the ossicles, which extend and transmit the vibrations from the eardrum to the inner ear. The internal ear consists of the cochlea, that is answerable for changing the vibrations from the ossicles into electric signals which can be despatched to the mind.

The cochlea is a spiral-fashioned shape that includes lots of tiny hair cells. These hair cells are responsible for converting the mechanical vibrations of sound waves into electric indicators that can be despatched to the mind. Different hair cells are activated with the resource of specific frequencies of sound, with better frequencies activating hair cells close to the lowest of the cochlea and reduce frequencies activating hair cells near the apex.

The thoughts then interprets the ones electric powered indicators as sound, permitting us to understand and choose out exceptional varieties of sounds. Our capacity to understand sound is prompted through pretty pretty more than a few of things, which includes the extent, pitch, and timbre of the sound.

Overall, the human ear is a complex and complex organ that lets in us to understand the wealthy and varied global of sound round us. By statistics how the ear perceives sound, we will higher recognize and enjoy the nuances and subtleties of numerous varieties of track and audio recordings.

The basics of sound waves

Sound is a shape of strength that travels thru the air inside the form of waves. These waves are created via vibrations, together with those produced with the beneficial aid of a musical tool or the human voice.

Sound waves can be characterized via severa specific houses, at the side of frequency, amplitude, and wavelength. Frequency refers back to the variety of cycles, or vibrations, that arise in a second and is measured in hertz (Hz). The higher the frequency, the higher the pitch of the sound.

Amplitude, alternatively, refers back to the electricity of the sound wave and is measured in decibels (dB). The extra the amplitude, the louder the sound. Finally, wavelength refers to the distance among peaks or troughs in a legitimate wave and is measured in meters.

When sound waves are recorded onto a vinyl record, the sound is converted proper into a bodily example of the sound wave. This is completed by manner of the use of a transducer, which converts the sound wave into an electrical sign that can be used to create a bodily groove on the vinyl.

When the needle of a report player is placed into the groove of a vinyl record, it vibrates in response to the bodily form of the groove. These vibrations are then transformed again into an electrical sign that may be amplified and played thru audio gadget.

The bodily nature of vinyl records method that they are able to accurately reproduce the precise nuances and subtleties of a legitimate wave. This is because the grooves on a vinyl report are an unique instance of the sound wave that changed into recorded onto it.

Vinyl records are regularly revered for their potential to efficaciously reproduce the specific nuances and subtleties of sound. This is because of the physical nature of vinyl records, which permits for a more committed instance of the unique sound wave than digital audio formats.

When sound is recorded onto a vinyl document, it is bodily etched into the floor of the vinyl as a series of grooves. The shape of these grooves corresponds straight away to the form of the unique sound wave, shooting every detail of the sound because it end up to begin with produced. This manner that the bodily grooves on a vinyl document are an particular example of the actual sound wave, and that's what permits for the trustworthy replica of the sound.

When a file player needle is placed into the groove of a vinyl report, it vibrates in reaction to the bodily shape of the groove. These vibrations are then transformed once more into an electrical signal that may be amplified and completed via audio device. Because the grooves on a vinyl record are an actual illustration of the unique sound wave, this means that that the sound reproduced via the usage of a vinyl document is a completely dedicated instance of the real recording.

Digital audio formats like MP3s, instead, depend upon a device known as "sampling" to reproduce sound. This involves taking snapshots of the sound wave at ordinary durations, and then the usage of those snapshots to recreate the sound. While virtual audio formats can provide tremendous sound, they are in the long run confined by using the choice of the virtual sign. This manner that they will no longer be able to seize all the specific nuances and subtleties of the specific sound in the same manner that vinyl information can.

Overall, the physical nature of vinyl statistics way that they're capable of provide a more correct and sincere replica of sound than virtual audio formats. This is why many audiophiles and tune fanatics preserve to choose the nice and cozy, natural sound of vinyl information, regardless of the benefit and reputation of virtual codecs.

While virtual audio codecs like MP3s can provide extremely good sound, they'll be ultimately restricted thru the resolution of the virtual sign.

Frequency, amplitude, and wavelength

Lin turntables famously carved a brand defining niche for themselves via coining the phrase PRaT approximately their turntables. They called it Pace, Rhythm and Timing measuring 3 key components of the sound their turntables reproduced.

So, allow's talk approximately that within the context of Frequency, amplitude, and wavelength.

Frequency, amplitude, and wavelength are three key factors of a valid wave that decide how it's miles perceived via the human ear.

Frequency refers lower back to the quantity of instances a legitimate wave oscillates backward and forward consistent with 2nd, measured in Hertz (Hz). A higher frequency corresponds to a higher pitch, even as a lower frequency corresponds to a decrease pitch. For example, a immoderate-pitched whistle should have a better frequency than a low-pitched bass guitar have a look at.

Amplitude refers to the maximum displacement of a valid wave from its resting feature, measured in decibels (dB). A huge amplitude corresponds to a louder sound, at the identical time as a smaller amplitude corresponds to a softer sound. For example, a thunderclap can also have a bigger amplitude than a whisper.

Wavelength refers to the gap amongst consecutive factors on a valid wave which might be in phase or have the same diploma of oscillation. Wavelength is measured in meters (m) or first rate distance units. A shorter wavelength corresponds to a better frequency, whilst an extended wavelength corresponds to a decrease frequency. For instance, the wavelengths of the immoderate-pitched sound produced thru manner of a whistle are shorter than the wavelengths of the low-pitched sound produced by using way of a bass guitar.

In precis, frequency, amplitude, and wavelength are 3 important characteristics of a legitimate wave that determine how it's far perceived via the human ear. By information these developments, we're capable of better appreciate and take a look at particular varieties of track and audio recordings.

Or to position it some other way vinyl records create frequency, amplitude, and wavelength by manner of using a physical grooved floor to file sound. The grooves on a vinyl document correspond to the vibrations of the precise sound wave, with the amplitude and frequency of the sound wave being encoded as variations in the depth and width of the groove.

When a stylus (needle) is located within the groove of a vinyl report and the document is spun, the stylus is vibrated via the variations within the groove. This vibration is then amplified and transformed into an electrical sign that may be performed thru a speaker, ensuing in sound that replicates the precise audio recording.

The bodily nature of vinyl statistics manner that they may accurately reproduce the ideal nuances and subtleties of a valid wave. The intensity and width of the groove on a vinyl record can seize the subtle differences in amplitude and frequency of a legitimate wave, resulting in a extra unique and correct duplicate of the real audio recording. Additionally, the analogue nature of vinyl statistics affords a warmer and further herbal sound in contrast to digital audio formats, that might frequently sound cold and sterile.

Vinyl Records - The Basics

Chapter 10: Structure Of A Vinyl File

It's not unusual understanding that vinyl records encompass a circular disc manufactured from polyvinyl chloride (PVC) plastic with a diameter of both 7 inches (for singles) or 12 inches (for albums or LPs). The ground of the report is grooved in a spiral pattern that starts offevolved at the outer edge and ends on the centre.

But did that Polyvinyl chloride (PVC) is a artificial polymer, due to this it is a form of plastic that is made through chemically combining specific molecules. PVC is made via polymerizing vinyl chloride monomers, which results in a thermoplastic cloth that can be ordinary and moulded into hundreds of bureaucracy.

PVC is notably carried out in diverse programs, along side creation, vehicle, electric powered, and medical industries, due to its sturdiness, versatility, and affordability. PVC is idea for being evidence towards chemical substances, weathering, and UV radiation, which makes it appropriate for outdoor use.

Overall, PVC is a broadly used synthetic polymer this is recognized for its sturdiness and flexibility, but moreover has functionality environmental and fitness worries related to its use.

Those groovy, grooves

The grooves at the document are wherein the audio statistics is saved. The intensity and width of the grooves correspond to the amplitude and frequency of the sound wave, respectively. A record participant makes use of a stylus or needle to trace the grooves due to the fact the document spins. The vibrations of the stylus are then transformed into an electrical sign, that is amplified and accomplished thru audio machine to deliver sound.

The vinyl cloth used to make information is a sort of plastic this is each long lasting and bendy. To make a file, PVC pellets are melted and pressed proper right into a flat disc known as a "biscuit". The biscuit is then positioned onto a stamping device, wherein a hydraulic press is used to stamp the grooves onto the floor of the report.

After the grooves are stamped onto the report, it undergoes a chain of strategies to refine and enhance it. This consists of trimming the edges of the file, smoothing out the ground, and applying a defensive coating to save you placed on and tear.

The label of a vinyl record is commonly positioned inside the centre of the disc and includes statistics approximately the artist, album become aware of, and tune listing. Some statistics may additionally moreover characteristic additional art work or information on the quilt or sleeve of the record.

In precis, the form of a vinyl record includes a spherical disc made from PVC plastic with grooves at the ground wherein audio information is saved. A stylus is used to examine the grooves and convert the vibrations into an electrical signal, it really is then carried out through audio device to supply sound. The vinyl material used to make statistics is long lasting and bendy, and information go through severa techniques to refine and beautify them in advance than they may be prepared for playback.

How vinyl records artwork

Vinyl records generate sound through using a mechanical technique to translate the grooves at the document into an electrical signal that may be amplified and played thru audio machine. Here's a step-through-step breakdown of methods vinyl facts paintings:

The vinyl document is positioned onto a turntable, which spins the document at a regular speed (usually 33 1/3 or 45 RPM).

A stylus, or needle, is positioned onto the ground of the spinning report. The stylus is hooked up onto the save you of a tonearm, this is related to a cartridge that consists of a tiny magnet and a coil of wire.

As the stylus actions alongside the grooves on the report, it vibrates in reaction to the modifications inside the groove intensity and form. These vibrations are then picked up through the magnet inside the cartridge, which generates an electrical sign in the coil of cord.

The electric powered powered signal is despatched thru the tonearm and proper proper right into a phono preamp, which amplifies and equalizes the signal. This is crucial due to the reality the signal coming off a vinyl document is a lot weaker and plenty less regular than tremendous audio assets.

The amplified signal is then despatched to an audio amplifier, which similarly amplifies the sign and sends it to audio device or headphones.

The audio device or headphones then convert the electrical signal again into sound waves, which can be heard via the listener.

Overall, vinyl information paintings by using the use of the usage of a mechanical technique to translate the bodily grooves at the file into an electrical signal that can be amplified and done decrease lower back as sound. This technique allows for a completely unique and heat sound tremendous that many track fanatics discover appealing.

The vinyl groove and its bodily traits

Vinyl facts have physical tendencies that contribute to their precise sound great. One of the most important of these characteristics is the report's groove. The groove on a vinyl record is a bodily example of the sound wave that changed into recorded onto the file. The width, depth, and spacing of the grooves are all carefully crafted to efficiently reproduce the real sound.

The grooves on a vinyl record are created at some point of the studying technique, wherein a recording engineer cuts the sound wave onto a smooth disc the usage of a lathe. This machine entails the use of a stylus to physical etch the groove into the report, with the width and depth of the groove much like the frequency and amplitude of the sound wave.

Vinyl data furthermore produce other physical characteristics that could have an effect on their sound top notch, together with the thickness and weight of the document. Generally, thicker and heavier records usually usually have a tendency to have better sound incredible than thinner and lighter statistics due to the truth they're lots much less prone to warping and vibration.

Vinyl records are physical devices that may be affected by a selection of things, which in turn will have an effect on their sound first rate. Some of the essential problem physical developments that may have an impact at the sound notable of a vinyl file encompass:

Thickness and weight: The thickness and weight of a vinyl file can also want to have a significant effect on its sound quality. Thicker and heavier data tend to have higher sound exceptional because of the truth they'll be an awful lot much less vulnerable to warping and vibration. A thicker report can also preserve a deeper groove, taking into account a extra positive and accurate sound replica.

Shape: The shape of a vinyl report can also have an effect on its sound notable. Records that aren't flawlessly flat or have uneven edges can purpose the stylus to bypass or soar, ensuing in distortion or skipping within the audio.

Surface noise: Surface noise is the sound of the stylus journeying over the floor of the report and can be because of dirt, scratches, or unique imperfections at the floor of the record. While a few diploma of surface noise is inevitable, excessive floor noise can extensively degrade the sound high-quality of a vinyl record.

Warping: Vinyl facts can warp because of publicity to heat or pressure, that could reason the stylus to song inconsistently and

Chapter 11: The Process Of Vinyl Record Production

Recording, mixing, and mastering

The manner of manufacturing a vinyl report begins offevolved with the recording technique, which includes taking pictures and storing an audio sign onto a draw close tape or virtual layout. The amazing of the precise recording will considerably have an effect at the very last sound exceptional of the vinyl record.

During the studying process, the audio signal is processed and equalized to optimize it for vinyl playback. This consists of adjusting the quantity ranges, frequency reaction, and dynamic range to make sure that the tune sounds balanced and easy on a vinyl report. A expert getting to know engineer could make a considerable difference in the very last sound extremely good of a vinyl report.

The mastering technique is a important step in generating notable vinyl data. Mastering is the very last diploma of audio manufacturing earlier than manufacturing and involves the education of the final mixture for optimum playback on vinyl. The getting to know engineer need to stability the tonal balance and dynamic variety of the recording and prepare it for maximum wonderful playback on vinyl.

The analyzing method starts with a switch of the final combination to a exceptional tape device. The tape gadget is then used to make EQ changes and dynamic variety processing. This degree is essential to ensure the recording sounds as incredible as feasible on vinyl. The studying engineer additionally creates the final collection of the tracks, which includes figuring out the strolling order of songs at the report.

Once the analyzing method is whole, a lacquer is reduce. The lacquer is an aluminum disc covered in a layer of nitrocellulose lacquer, this is used to create the master duplicate of the recording. The getting to know engineer cuts the grooves into the lacquer with a reducing head that follows the movements of the particular aggregate. The reducing head produces the grooves with a view to later be pressed onto the vinyl.

The lacquer is then sent to a urgent plant, in which it is used to create a steel stamper. The stamper is a terrible of the lacquer and is used to press the grooves into the vinyl. The stamper is hooked up proper into a vinyl pressing system, and the vinyl pellets are heated and pressed into the grooves of the stamper. The vinyl is then cooled and eliminated from the gadget, and the manner is repeated to create more copies of the document.

The lacquer disc is an crucial issue within the way of creating vinyl facts. The lacquer disc is the first step in developing the draw close duplicate of the vinyl document. It is a skinny aluminium disc blanketed with a unique lacquer this is gentle and malleable. The lacquer disc is set up onto a lathe, that is a machine used to cut the grooves into the disc.

The grooves are reduce into the lacquer disc using a reducing stylus, this is linked to the lathe. The lowering stylus vibrates to and fro in response to the sound waves from the recording, and this vibration is used to reduce the grooves into the disc.

The lacquer disc is a touchy and fragile element, and it may simplest be played a constrained form of instances earlier than it starts offevolved offevolved to degrade. After the grooves had been reduce into the lacquer disc, it's far protected with a thin layer of metallic, generally nickel, to create a steel terrible referred to as a "metal mother."

The metal mother is then used to create a metallic stamper, this is the very last step within the technique of making a vinyl record. The stamper is used to press the grooves into the vinyl, growing the final product so one can be bought to purchasers. The lacquer disc is not used yet again inside the approach of making vinyl facts, as it's far too sensitive and may most effective be used a limited quantity of instances.

The analyzing gadget is important to making sure that the very last vinyl record sounds as properly as feasible. The lacquer lowering technique is step one in developing a physical duplicate of the recording, and the stamper is used to create the grooves at the vinyl. Each step inside the technique is important to developing notable vinyl statistics.

Once the studying way is entire, the audio is transferred to a lacquer disc, which serves because the template for growing the vinyl document. The lacquer disc is included with a thin layer of metal, normally silver or nickel, after which electroplated with a layer of nickel. This creates a negative photo of the audio, that is then used to create the stamper so that you can be used to press the vinyl information.

The stamper used to press vinyl facts is a vital issue inside the manufacturing way. It is largely a metal plate with a horrible photo of the audio recording etched onto its floor. The stamper is created by the use of electroplating a layer of nickel onto a lacquer disc that has been included with a thin layer of metallic.

During the urgent device, the stamper is placed proper into a hydraulic press, which applies warmth and pressure to the vinyl material. The grooves are pressed into the vinyl, and the greater cloth is trimmed away. The stop result is a finished vinyl report with grooves that correspond to the audio sign etched into the stamper.

The stamper must be carefully synthetic to make certain that the grooves are well aligned and that the depth and shape of the grooves are regular in the course of the disc. Any defects or irregularities inside the stamper may be transferred to the vinyl record, ensuing in audible distortion or noise.

Because the stamper is a terrible photograph of the audio recording, it may best be used to provide a finite variety of vinyl statistics. As the stamper is used to press more statistics, it regularly wears down and the super of the finished records can also moreover moreover degrade. For this motive, many report urgent plants restrict the amount of data that may be pressed from a single stamper.

In order to hold remarkable records, the stamper have to be carefully wiped smooth and maintained amongst pressings. Any dust or debris that accumulates on the stamper can be transferred to the vinyl report, resulting in audible noise or distortion. Therefore, the stamper is a essential component inside the vinyl record manufacturing method and must be carefully crafted and maintained to provide extremely good facts.

The stamper is located right right into a hydraulic press, which applies warm temperature and stress to the vinyl cloth. The grooves are pressed into the vinyl, and the more material is trimmed away. The completed document is then inspected for defects, cleaned, and packaged for distribution.

Throughout the gadget, interest to detail is important to make certain that the completed product sounds as appropriate as viable. Any flaws inside the recording, gaining knowledge of, or urgent technique can bring about degraded sound pleasant, so it is crucial to have expert professionals coping with every step of the method.

The function of the reducing lathe in vinyl production

The cutting lathe is a key element in the vinyl report manufacturing method. It is used in the analyzing method, in which the final audio recording is transferred to a lacquer disc, with a purpose to then be used to create the metallic stampers for urgent the vinyl statistics.

The reducing lathe operates thru using receiving an electrical sign from the audio supply and changing it right into a bodily groove at the lacquer disc. The lathe is geared up with a slicing head that homes a stylus, which etches the groove onto the ground of the lacquer disc.

The reducing head is designed to move from side to side in reaction to the electric sign it gets, cutting a groove that successfully shows the sound wave. The depth, width, and shape of the groove are decided with the aid of the amplitude, frequency, and wavelength of the audio signal.

The reducing lathe operator has a important role inside the approach, adjusting the settings on the lathe to make sure that the groove is cut to the right depth and width, and that the overall sound exceptional is preserved. The operator will use specialized device to degree and display the groove as it's far being lessen, adjusting as critical.

The first-rate of the slicing lathe and the know-how of the operator could have a massive impact at the very last sound top notch of the vinyl document. A well-maintained and calibrated lowering lathe, operated via an professional technician, can produce a document that faithfully reproduces the authentic audio deliver with superb fidelity and readability.

The manner of urgent vinyl statistics

The vinyl material used for pressing is made thru using heating and melting polyvinyl chloride (PVC) pellets, which is probably then not unusual into small "biscuits." These biscuits are then cooled and trimmed to the right size for pressing.

The vinyl pressing procedure begins with the aid of heating the biscuits to a temperature of round one hundred sixty-100 seventy degrees Celsius, which makes the PVC fabric smooth and pliable. A stamper is then positioned on top of the biscuit and hydraulic strain is implemented, forcing the material to take on the form of the groove in the stamper. The resulting disc is then cooled and trimmed to length.

Once the vinyl discs are pressed, they go through an extensive amazing manipulate device to make certain that they meet the favored specs for sound wonderful and durability.

The pressing manner can be a sensitive stability among art work and technological expertise, with many factors which incorporates temperature, pressure, and timing affecting the final sound notable of the vinyl report. Skilled engineers and technicians play a crucial feature in ensuring that the very last product meets the immoderate standards predicted through audiophiles and tune fans.

Chapter 12: Groove Geometry

The importance of groove geometry

The groove geometry of a vinyl document plays a crucial function in figuring out the sound extraordinary of the final product. The groove geometry refers to the bodily shape and dimensions of the grooves that are lessen into the lacquer maintain near disc and in the end pressed into the vinyl report.

The depth and width of the grooves, in addition to the spacing among them, are cautiously designed to capture and reproduce the nuances of the sound being recorded. The reducing engineer uses specialized software application and device to create the groove geometry, taking into account factors which encompass the frequency reaction of the reducing head, the stylus tip duration, and the trends of the vinyl cloth itself.

One vital aspect of groove geometry is the angle at which the stylus enters the groove. This perspective, referred to as the "reducing angle," affects the amount of physical stress that the stylus exerts at the vinyl cloth, similarly to the amount of distortion that can be delivered into the sound signal. Cutting engineers want to carefully balance the ones factors to attain the favored sound notable.

Another vital issue of groove geometry is the spacing some of the grooves. This spacing is called the "music pitch" and is determined via the bodily boundaries of the reducing device and the popular period of the very last file. A narrower music pitch allows for longer recording times but also can bring about lower sound incredible, on the equal time as a far wider track pitch lets in for higher sound high-quality however shorter recording instances.

Overall, the groove geometry of a vinyl file is a vital factor of the manufacturing machine that requires cautious interest to detail and technical data. When finished efficaciously, it could result in a first rate sound recording that as it should be captures the nuances and subtleties of the correct normal overall performance.

How groove intensity, width, and spacing have an effect on sound

Groove intensity, width, and spacing are essential bodily dispositions that could have a huge impact at the sound brilliant of a vinyl record. These factors affect the quantity and form of the vibrations which may be captured thru the stylus as it travels thru the groove, which in turn determines the electric signal that is despatched to the amplifier and in the end to the audio gadget.

Groove depth refers to the distance among the very incredible factor of the groove and the bottom of the groove. A deeper groove can capture more sound information and offer higher bass response, but it may moreover growth the hazard of the stylus jumping out of the groove and causing distortion. Conversely, a shallower groove might also sacrifice some low-prevent frequency reaction but can offer extra regular clarity and decrease the opportunity of stylus skipping.

Groove width is the distance a number of the walls of the groove. A wider groove can accommodate more sound statistics and provide better dynamics, however it may additionally increase the chance of crosstalk among adjoining grooves, that could bring about distortion and absence of stereo separation. Narrower grooves can reduce the threat of crosstalk but might not capture as a whole lot sound information and may bring about a weaker common sound.

Groove spacing refers to the distance amongst each successive revolution of the groove. This impacts the amount of playing time that may be accommodated on a single aspect of a vinyl file, with wider groove spacing allowing for longer gambling instances. However, wider groove spacing can also reduce the general sound high-quality via increasing the threat of crosstalk and reducing the amount of records that may be captured in the groove.

Overall, carrying out the quality stability among groove depth, width, and spacing is crucial to producing a first-rate vinyl file with high-quality sound replica. This requires careful interest of the recording, reading, and slicing techniques, in addition to the selection of amazing materials and device.

Different varieties of groove shapes

Vinyl data may also moreover have severa styles of groove shapes, that could have an effect on the sound exquisite and common listening experience. Here are some of the most commonplace groove shapes:

Standard groove: This is the most common type of groove shape, and it's miles used for maximum facts. The popular groove has a V-shape, with the partitions of the groove forming an attitude of round 90 levels.

The vast groove form for vinyl records is the most used and properly-installation shape. This groove shape has a V-unique move-segment and is symmetrical in format. It capabilities partitions that slope towards each other to form a thing at the lowest of the groove, developing a sharp facet that permits produce a extra particular and described sound.

The massive groove form is taken into consideration a extremely good all-round shape that gives a stability between sound tremendous and durability. The V-shape of the groove permits to reduce ground noise, at the same time as the pointy edges offer excellent monitoring capability and excessive-fidelity playback.

One gain of the standard groove shape is that it is highly clean and fee-powerful to deliver, making it a famous desire for record manufacturers. Additionally, it really works properly with a whole lot of musical styles and genres, making it a flexible opportunity for each audiophiles and casual listeners.

However, there are a few drawbacks to the same vintage groove form. One problem is that it is able to be prone to groove wear through the years, which could bring about a lack of fidelity and readability. Additionally, due to the fact the groove is symmetrical, there is a lot tons less area to be had for the audio sign, that could restrict the dynamic form of the recording.

Despite those obstacles, the standard groove form remains the maximum used shape for vinyl information and is extensively recognized as a reliable and powerful method of reproducing sound.

Microgroove: Microgroove information were added inside the Fifties as a manner to growth the gambling time of information while maintaining sound best. Microgroove statistics have narrower grooves and a better range of grooves regular with inch, which allows for greater tune to be packed onto a single element of the report. However, the narrower groove can purpose a lack of excessive-frequency statistics.

It emerge as introduced to growth the amount of tune that might healthy on a single report, while maintaining sound high-quality. The Microgroove form has a smaller groove width of approximately 0.001 inches (25 microns) and a shallower groove depth of about zero.00006 inches (1.Five microns). The groove spacing is also narrower, bearing in thoughts extra grooves regular with inch. This allows for as an awful lot as 22 mins of audio consistent with side on a 12-inch vinyl record, in evaluation to the four-five minutes consistent with thing on a 78 rpm shellac disc.

The smaller groove length requires a greater specific cutting manner, as any deviation from the desired form ought to have a greater impact on sound exceptional. However, the shallower groove intensity additionally way that the stylus might now not want to song as deeply into the groove, decreasing the danger of distortion and placed on on each the record and stylus.

Overall, the Microgroove form has allowed for longer gambling times, advanced sound first-rate, and reduced put on and tear on each records and playback device. It has end up the corporation stylish for vinyl file manufacturing.

Elliptical groove: This groove form is regularly used on audiophile facts, which are designed for brilliant playback systems. Elliptical grooves have a greater gradual slope than stylish grooves, which lets in for extra real sound duplicate.

The elliptical groove form is a type of groove used on some vinyl information. The form of the groove is elliptical, which means it is elongated and now not perfectly round like a conventional groove. This form is designed to permit for better tracking of the stylus, or needle, thru the grooves.

The elliptical shape gives a larger touch location for the stylus, allowing it to hint the groove and select up greater element inside the tune extra correctly. This can result in advanced immoderate frequency reaction and everyday sound brilliant.

One disadvantage of the elliptical form is that it can reason more put on and tear on every the document and the stylus. The elongated form can placed extra pressure on a smaller region of the groove, important to faster wear and feasible distortion over the years. As a give up stop result, elliptical stylus suggestions want to be well maintained and modified often to ensure the superb feasible sound exquisite and longevity of the record.

Bi-radial groove: This groove form is used on quadraphonic information, which were designed to be executed on 4-channel sound structures. Bi-radial grooves are designed to accommodate each lateral and vertical stylus motion.

The bi-radial groove form for vinyl facts is a variation on the elliptical groove shape that modified into introduced by way of Neumann in the Seventies. The bi-radial form grow to be designed to further lessen distortion and boom the overall frequency response of the report.

In a bi-radial groove, the curvature of the groove changes as it movements in the route of the centre of the file. The outer part of the groove is shallower and wider, even as the internal aspect is deeper and narrower. This form permits for a higher stage of precision inside the modulation of the groove, ensuing in a extra accurate instance of the recorded sound.

The bi-radial shape moreover lets in to lessen the quantity of monitoring mistakes, which occurs whilst the stylus deviates from the intended course because of variations inside the groove form. This is performed with the useful aid of developing a greater sluggish transition a few of the outer and internal grooves, lowering the abrupt modifications in curvature that would motive tracking errors.

The bi-radial groove shape is a massive improvement over preceding designs and has been extensively followed in cutting-edge vinyl manufacturing. It is a testament to the continuing efforts to decorate the fidelity and accuracy of vinyl records as a medium for audio playback.

W-original groove: This groove shape have grow to be used on some early stereo information. W-authentic grooves have partitions on each aspect of the groove, which lets in for greater actual stereo separation.

The W-fashioned groove is a shape of groove shape carried out in vinyl information, and it is also called a zig-zag or multi-ridge groove. This groove form is designed to increase the amount of tune that may be packed onto a unmarried record without sacrificing sound super.

The W-long-established groove is carried out through way of reducing a series of ridges into the vinyl at an attitude, growing a chain of zig-zags across the floor of the report. The ridges are spaced extra intently collectively near the centre of the file and steadily unfold apart in the direction of the outer edge, allowing extra tune to be in shape onto the disc even as keeping a regular groove intensity.

One advantage of the W-regular groove is that it allows for longer gambling instances on every element of the report. Since the grooves are closer together near the centre of the disc, extra music can be packed onto the innermost tracks. This lets in for longer playing times without having to lessen the extent or wonderful of the song.

However, there are a few potential downsides to the W-shaped groove. One is that the ridges can create greater noise or distortion, specially on older or much less precise turntables. Additionally, the nearer spacing of the grooves close to the centre of the file should make it greater difficult for the stylus to appropriately song the groove, potentially causing skipping or one-of-a-kind playback problems.

Chapter 13: Cartridge Styles And Types

The type of stylus decided on is vital inside the reproduction wonderful of the vinyl. As there are such some of potential groove profiles used to create vinyl facts, the choice of stylus is continuously going to be a 'awesome-healthful' compromise. So, allow's take a look at the specialists and cons of the maximum well-known, however earlier than that assessment the fundamentals, that's the position of the stylus in playback.

As it's the first-class thing in direct touch with the music supply, many audiophiles would possibly argue that the cartridge and particularly the stylus is the maximum important a part of howdy-fi tool.

The characteristic of the cartridge in playback

The cartridge is a important detail in the playback of vinyl records. It is answerable for converting the mechanical power of the stylus because it tracks the grooves of the document into an electrical signal that can be amplified and reproduced as sound. The cartridge includes a frame that houses the stylus, cantilever, and magnet, in addition to a fixed of coils that generate the electric signal.

The layout and great of the cartridge may have a huge impact at the sound high-quality of the playback. Some of the elements that may have an impact on the general overall performance of a cartridge embody the mass of the moving elements, the shape and fabric of the cantilever, the power of the magnet, and the brilliant of the coils.

Moving Magnet (MM) cartridges are the maximum common shape of cartridge and are stated for his or her excessive output and ease of use. Moving Coil (MC) cartridges, as an alternative, have a decrease output but are appeared for his or her superior sound awesome due to their decrease mass and higher compliance.

The choice of cartridge can also depend upon the shape of tune being accomplished and personal alternatives. For example, a cartridge that is good for reproducing classical track won't be brilliant for playing rock or hip-hop. It is critical to pick out a cartridge this is properly-first-class for the music being achieved and that gives the popular sound first-class.

Different varieties of cartridges and their traits

There are principal varieties of cartridges used for vinyl file playback: Moving Magnet (MM) and Moving Coil (MC).

Moving Magnet (MM) cartridges are the most common kind and are quite inexpensive as compared to Moving Coil cartridges. They have a higher output voltage and are like minded with maximum phono preamps. They furthermore have a tendency to have a hotter sound because of their decrease mass and much less touchy nature. However, their better compliance can result in much a good deal less accurate monitoring and masses much less element retrieval.

Moving Coil (MC) cartridges are commonly extra high priced and feature a lower output voltage in comparison to MM cartridges. They have a lower compliance and are more touchy, resulting in extra correct tracking and better detail retrieval. They are also known for his or her speedy brief reaction and flat frequency reaction. However, they require a specialised phono preamp that could offer the important benefit and impedance matching, that may upload to the general price.

Another form of cartridge is the Moving Iron (MI) cartridge, which uses a hard and fast magnet and movable iron cantilever. These cartridges are stated for their wide frequency response and low distortion but are a whole lot less commonplace and can be hard to find out replacements for.

There also are immoderate-output cartridges, which have a better output voltage and do now not require as plenty amplification from the phono preamp. Low-output cartridges, however, require greater amplification and are generally extra sensitive to noise and hum.

In addition, cartridges also can variety of their stylus type and shape, similarly to the substances used of their introduction, together with the cantilever, magnet, and coil cord. Different materials can have an effect on the sound extraordinary, with a few substances imparting a warmer or extra precise sound, even as others may be extra independent or apparent.

How to pick out out the right cartridge to your turntable

When choosing the proper cartridge in your turntable, there are numerous elements to take into account:

Compatibility: The cartridge must be properly perfect along side your turntable's tonearm. Different turntables have specific tonearm mass and impedance requirements, and cartridges are designed to artwork optimally inside particular ranges. Be certain to test your turntable's specs and go to the manufacturer or an professional to make certain compatibility.

Budget: Cartridges can range from very low-price to very high-priced. It's vital to decide your finances and then take a look at the alternatives available indoors that variety.

Sound terrific: Different cartridges should have awesome sound trends, so it is vital to pick one that fits your alternatives. Some cartridges are recognized for having a heat, mellow sound, whilst others have a shiny, special sound. Reading evaluations and taking note of sample tracks will let you decide which cartridge will produce the sound you are looking for.

Type of tune: Some cartridges are better ideal for wonderful varieties of track. For instance, if you pay attention regularly to classical or jazz track, a cartridge with a excessive level of element and accuracy can be more vital than one with masses of bass. On the other hand, if you pay attention to numerous rock or digital music, a cartridge with a strong bass response can be greater crucial.

Upgradability: Some cartridges may be upgraded with a better stylus or possibly a better frame. If you decided you can want to beautify your cartridge within the future, search for one this is designed to be upgraded.

Tracking stress: Different cartridges require wonderful tracking forces, which could have an effect at the lifespan of your facts and the fantastic of the sound. Be effective to pick a cartridge that is nicely perfect with the tracking stress variety endorsed thru your turntable's manufacturer.

Choosing the right cartridge to your turntable calls for cautious consideration of severa factors. By doing all your research and consulting with experts, you could discover a cartridge for you to produce the sound awesome you're looking for and be well matched together together with your turntable for destiny years.

The All Important Stylus

The function of the stylus in playback

The stylus, moreover referred to as the needle, is a small, diamond-tipped device that sits in the groove of a vinyl file in the end of playback. As the record spins, the stylus vibrates up and down in response to the variations within the groove, growing an electrical sign that is despatched to the amplifier after which to the audio machine or headphones.

The form and top notch of the stylus are critical in figuring out the sound first-class of the playback. The stylus should be matched to the right groove shape and length of the document, in addition to to the tracking pressure of the turntable's tonearm. Too a good buy or too little tracking pressure can motive distortion, skipping, or even everlasting harm to the record and the stylus.

The form and brilliant of the stylus are important in identifying the sound extremely good of the playback due to the reality the stylus is chargeable for tracking the grooves at the vinyl file and translating the physical actions of the grooves into an electrical sign that can be amplified and heard via audio system. As the stylus tracks the grooves, it research mechanical forces and vibrations which could motive distortion, noise, and positioned on and tear on the stylus and the vinyl record itself.

The shape of the stylus tip determines the quantity of touch it has with the groove walls and the way nicely it can correctly song the excessive-frequency additives of the audio signal. A well-designed stylus might also have a easy, tapered profile that reduces floor noise and minimizes positioned on on the report. The splendid of the substances used to make the stylus is also essential, as tougher materials like diamond or sapphire are greater long lasting and might preserve their shape for longer durations of time, ensuing in better sound incredible over the lifespan of the stylus.

The form and wonderful of the stylus are crucial due to the truth they without delay have an effect at the ability of the stylus to because it want to be track the grooves and reproduce the sound recorded at the vinyl report and can also impact the lifespan of each the stylus and the file itself.

What are the factors of a stylus cartridge?

A stylus cartridge consists of numerous elements that artwork together to supply sound from a vinyl report. These factors include:

Cantilever: The cantilever is a small, thin rod that holds the stylus at one surrender and the magnet or coil at the opposite cease. It acts as a suspension system for the stylus, permitting it to transport up and down because it tracks the record groove.

Stylus: The stylus, moreover called the needle, is a small, diamond-tipped piece that tracks the file groove and produces sound vibrations.

Magnet or coil: The magnet or coil is a small element that converts the mechanical vibrations of the stylus into electric powered indicators.

Body: The frame of the cartridge is typically made from plastic or metal and houses the cantilever, stylus, and magnet or coil.

Terminals: The terminals are metallic contacts that allow the cartridge to be associated with a tonearm or preamp.

Shielding: The protective allows to save you interference from unique virtual additives and electromagnetic fields.

Suspension: The suspension gadget holds the cantilever in place and permits it to transport freely, even as additionally presenting balance and accuracy.

All of those elements paintings collectively to supply the sound that is heard at the identical time as playing a vinyl report.

Chapter 14: Different Sorts Of Cartridges

There are numerous styles of stylus cartridges, the most popular are Moving Magnet and Moving Coil. Below is a greater exhaustive listing of types of cartridges, they encompass:

Moving Magnet (MM) Cartridges: These are the maximum common shape of cartridge and feature a magnet related to the surrender of the cantilever that movements amongst coils to generate a sign.

Moving Coil (MC) Cartridges: These cartridges have a coil related to the give up of the cantilever that actions amongst magnets to generate a signal.

Moving Iron (MI) Cartridges: In the ones cartridges, a small piece of iron is established to the stop of the cantilever and actions between two magnets to generate a signal.

Ceramic Cartridges: These cartridges use a ceramic piezoelectric detail to generate a sign.

Optical Cartridges: These cartridges use a mild beam to find vibrations within the stylus, which might be then transformed into an electrical sign.

Strain Gauge Cartridges: These cartridges use a strain gauge to discover vibrations inside the stylus, which may be then transformed into an electrical sign.

The blessings and disadvantages of MC and MM cartridges

Moving Magnet (MM) cartridges are a famous form of stylus cartridge utilized in vinyl playback. Some of the specialists and cons of MM cartridges are:

Pros:

- Good sound amazing for the rate

- High output voltage, due to this they're nicely acceptable with a far wider kind of preamps and amplifiers

- Replaceable stylus, that is greater price-effective than replacing the whole cartridge

- Generally, more less expensive than Moving Coil (MC) cartridges

Cons:

- Less element and accuracy than MC cartridges

- Higher distortion tiers than MC cartridges

- Limited to a hard and fast coil impedance, which won't be suitable for all systems

- Generally, have a higher mass, that may affect monitoring capacity and purpose greater record placed on through the years.

Moving Coil (MC) cartridges are every other shape of phono cartridge utilized in vinyl playback structures. They fluctuate from

Moving Magnet (MM) cartridges of their layout, in which the stylus is installed to a moving coil that generates the electrical signal.

Pros of Moving Coil (MC) Cartridges:

- Better channel separation, which ends up in greater correct stereo imaging and soundstage

- More touchy to nuances and subtleties within the tune, resulting in a extra unique and realistic sound

- Lower transferring mass than MM cartridges, that would result in a quicker and further correct response to the groove modulations

- Generally, have a lower output impedance, that could paintings better with positive styles of phono preamps

Cons of Moving Coil (MC) Cartridges:

- Higher rate in contrast to Moving Magnet (MM) cartridges, that may restriction affordability for a few purchasers

- Lower output level in comparison to MM cartridges, which also can require a preamp with higher benefit to achieve exquisite enough extent

- Can be greater difficult to put in and installation because of their smaller length and similarly sensitive nature

- May require specialized device or modifications to art work well with excessive pleasant turntables and tonearms.

Whatever cartridge you use, it's miles vital to maintain the stylus smooth and loose from dirt and particles, as dust and excellent particles can reason harm to every the stylus and the file. Regular cleansing with a stylus brush or professional cleaning solution, as an example Stylus Clear can assist to maintain most efficient playback exquisite and amplify the lifespan of the stylus.

Different varieties of stylus shapes

Styli can are available wonderful shapes, collectively with conical, elliptical, and Shibata. Conical styli have a rounded tip and are a whole lot less highly-priced but can reason more positioned on and tear on the record and bring a terrific deal tons less element inside the sound. Elliptical styli have a more precise shape and might track the groove extra accurately, ensuing in higher sound amazing. Shibata styli are even more precise and can produce even higher sound but are more luxurious and require extra cautious coping with. Each shape impacts how the stylus interacts with the groove wall of the vinyl document and may have an effect on the sound exceptional of the playback.

Spherical stylus: This is the maximum essential and normally used stylus shape.

The tip of the stylus is an splendid sphere, which makes it durable and coffee-price. However, due to the reality the top is spherical, it most effective contacts a small part of the groove wall, which could bring about a lack of detail and accuracy in the sound reproduction.

The round stylus form is the maximum common sort of stylus determined in get entry to-level turntables and cartridges. It has a rounded shape that makes touch with the grooves at a unmarried trouble, permitting it to track a substantial kind of report surfaces.

Some of the specialists of a round stylus form embody:

1.   Durability: Spherical styli are less liable to harm from dust and dust particles and have a longer lifespan compared to distinct kinds of stylus shapes.

2.   Low tracking stress: Spherical styli require decrease tracking forces to hold

actual tracking, because of this that they're less probably to harm records or purpose excessive put on.

3. Wide compatibility: Spherical styli are well proper with a large fashion of statistics, which includes older and lower first-rate records that won't be playable with unique forms of styli.

However, there also are a few cons to take into account:

1. Sound outstanding: Spherical styli have a noticeably low contact region with the groove wall, resulting in an awful lot less superb sound duplicate as compared to particular stylus shapes.

2. Surface noise: Because they make contact with a smaller region of the groove wall, spherical styli can pick out out up extra surface noise and pops and clicks in comparison to distinctive shapes.

three.   Limited   excessive-frequency response: Spherical styli have limited excessive-frequency response.

Elliptical stylus: The elliptical stylus has a barely elongated shape, which lets in it to the touch a extra part of the groove wall. This consequences in superior sound exquisite and further detail inside the playback. Elliptical styli are more highly-priced than round styli, but are a famous desire for lots audiophiles.

Elliptical stylus shapes have several benefits and drawbacks:

Pros:

- Better immoderate-frequency response: Because the elliptical shape contacts the groove walls over a larger region, it could as it should be hint excessive-frequency waveforms that a round stylus should possibly bypass over.

- Reduced distortion: Elliptical styli have a smaller touch radius than spherical styli, which reduces distortion due to tracing errors and misalignments.

- Improved channel separation: The reduced contact vicinity of the elliptical stylus moreover way that it could greater correctly song the groove modulations in each channel, ensuing in progressed channel separation.

Cons:

- Increased groove wear: Because the elliptical form contacts the groove walls over a smaller area than a round shape, the expanded pressure on the groove can bring about faster wear.

- Higher fee: Elliptical styli are extra complicated to fabricate than spherical styli, and therefore will be inclined to be greater pricey.

- More sensitive to alignment: Elliptical styli require extra particular alignment than spherical styli to make sure crucial usual performance, which may be hard for a few turntable setups.

Microline stylus: Also known as a Shibata stylus, the microline stylus has a greater complex form that resembles a triangle with a flat pinnacle. This shape allows it to make even extra contact with the groove wall than the elliptical stylus, resulting in even greater detail and accuracy within the sound duplicate.

Microline styli are the most high-priced of the three types of stylus shapes, but are regularly taken into consideration the great desire for crucial audiophiles.

SAS stands for Shibata and is a type of stylus form utilized in excessive-prevent cartridges for gambling vinyl records. The Shibata shape emerge as first superior by using the use of Norio Shibata in the early Sixties and is characterised with the aid of a exceptionally polished, rectangular-cut diamond this is ground to a satisfactory thing.

The shape of the SAS stylus lets in it to music the groove walls of a document greater successfully than distinct types of stylus shapes. This is because of the reality the rectangular form of the diamond gives a bigger floor location for contact with the groove partitions, which reduces the amount of stress exerted on any single component of the groove. This, in flip, minimizes the quantity of wear and tear and tear at the file and reduces distortion and noise.